Mathematical Foundations of Big Data Analytics

Vladimir Shikhman • David Müller

Mathematical Foundations of Big Data Analytics

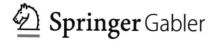 Springer Gabler

Vladimir Shikhman
Fakultät für Mathematik
Chemnitz University of Technology
Chemnitz, Germany

David Müller
Fakultät für Mathematik
Chemnitz University of Technology
Chemnitz, Germany

ISBN 978-3-662-62520-0   ISBN 978-3-662-62521-7   (eBook)
https://doi.org/10.1007/978-3-662-62521-7

This Springer Gabler imprint is published by the registered company Springer-Verlag GmbH, DE part of Springer Nature.
The registered company address is: Heidelberger Platz 3, 14197 Berlin, Germany

Preface

Subject

Big Data Analytics is a relatively modern field of data science that explores how available data sets can be broken down and analyzed in order to systematically glean insights and information from them. What is characteristic of Big Data Analytics is that the data to be processed is too large, too complex, too fast moving, too costly, or too weakly structured to be evaluated using manual and conventional processing methods. This is often referred to as the concept of the Five Vs:

- Volume defines the huge amount of data that is produced and stored by private and public companies, research institutions, and households.
- Variety reflects the diversity of data types and data sources, which include consumer profiles, social contacts, texts, images, videos, and even speech recordings.
- Velocity means the high speed with which the data is continually generated, analyzed, and reprocessed in order to support the underlying decision making.
- Validity is the guarantee of data quality, or alternatively, its authenticity and credibility, elements that often suffer due to possible measurement inaccuracies.
- Value comes from the cost-benefit analysis and denotes the positive impact generated for companies through systematic gathering and use of data in their business activities.

Altogether, the challenge of the Five Vs nowadays is to propose adequate concepts and algorithms aimed to capture, store, process, or harness data effectively.

Concept

In this textbook, basic mathematical models used in Big Data Analytics are presented and application-oriented references to relevant practical issues are made. Necessary mathematical tools are examined and applied to current problems of data analysis, such as brand loyalty, portfolio selection, credit investigation, quality control, product clustering, asset pricing, etc.—mainly, in an economic context. In addition, we discuss interdisciplinary applications to biology, linguistics, sociology, electrical engineering, computer science, and artificial intelligence, and our examples include DNA sequencing, topic extraction, community detection, compressed sensing, spam filtering, and chess engines. For the models, we make use of a wide range of mathematics—from basic disciplines of numerical linear algebra, statistics, and optimization to more specialized game, graph, and even complexity theories. By doing so, we cover all relevant techniques commonly used in Big Data Analytics. These correspond to the chapters on ranking, online learning, recommendation systems, classification, clustering, linear regression, sparse recovery, neural networks, and decision trees. The structure and size of the chapters are standardized throughout the textbook for students' and teachers' convenience.

Each chapter starts with a concrete practical problem whose primary aim is to motivate the study of a particular Big Data Analytics technique. Next, mathematical results follow—including important definitions, auxiliary statements, and conclusions arising. Case studies help to deepen the acquired knowledge by applying it in an interdisciplinary context. A case study consists of a description of step-by-step tasks accompanied by helpful hints. Exercises, as an indispensable part of self-study, serve to improve understanding of the underlying theory. Complete solutions for exercises can be consulted by the interested reader at the end of the textbook; for some of those that have to be numerically, we provide descriptions of algorithms in Python code as supplementary material.

Audience

The target group includes graduate students generally focused on Big Data Analytics—its mathematical foundations and relevant applications—in their studies. In the last few years, corresponding Master's programs, such as Data Engineering and Analytics, Computational and Data Science, Big Data and Business Analytics, Management and Data Science, Social and Economic Data Science, Data Analysis and Decision Science, Big Data Management, Data Science in Business and Economics, and Machine Learning, have multiplied in the world's best universities. Usually, such programs are organized by economists, mathematicians, computer scientists, and/or engineers, implying a certain heterogeneity of students' background and skill sets. We cater to this interdisciplinarity by carefully elaborating the mathematical foundations of Big Data Analytics and providing applications from nearly all important areas of research. The level of training required corresponds to standard courses in higher mathematics, regardless of whether they are aimed for economists, computer scientists, physicists, chemists, electrical or machine engineers. All of these professionals would gain from being acquainted with Big Data Analytics, as this domain is playing an increasingly important role within their own fields. Moreover, graduate students will also get insight into the area of data science, which is already a tool for considerable impacts and change in our society and will, hopefully, contribute to its further success in the future.

Acknowledgment

Our first thanks go to Friedrich Thießen and Peter Gluchowski who initiated the development of the course "Mathematical Foundations of Big Data Analytics" at the Chemnitz University of Technology in 2017. As those responsible for Master Finance and Master Business Intelligence & Analytics programs, respectively, they challenged us to design an interdisciplinary course on data analysis as a part of the corresponding curricula. Friedrich Thießen helped a lot in identifying practical questions and problems of particular economic interest connected to data science. The discussions with Peter Gluchowski on the structure of the course and material selection have proved to be crucial.

Second, we would like to thank our colleagues Oliver Ernst, Roland Herzog, Alois Pichler, and Martin Stoll from the Faculty of Mathematics at Chemnitz University of Technology. As early as 2018, we began to collaborate on a Master of Data Science program, where the teaching of "Mathematical Foundations of Big Data Analytics" could be successively tried and improved. These experiences—advanced by interesting conversations on the didactics of data science with those colleagues—resulted in the present textbook.

Third, we are grateful to Iris Ruhmann from Springer for giving us advice and support during the manuscript preparation. Her innovative perspective on how mathematical knowledge can be introduced into an interdisciplinary context was very motivating. We also would like to thank Greta Marino and Rory Sarkissian for carefully checking the parts of the manuscript. Last but not least, the comments from our students who pointed out typos and inaccuracies are also very much appreciated.

Chemnitz, Germany   Vladimir Shikhman
August 2020   David Müller

Contents

Ranking

We face *rankings* in our daily life rather often, e.g. consumer organizations rank products according to their qualities, scientists are ranked upon their publications, musicians aim for a top chart position, soccer teams compete for wins in order to climb up in the league table and so on. Thus, the central idea of a ranking is to arrange subjects in such a way that "better" subjects have higher positions. Obviously, most of the rankings are based on an intuitive order, e.g. more victories of a team lead to a higher place in the soccer league, the higher quality index should result in a more valuable ranking for consumption of goods and services. Apart from these examples, rankings can also be derived just out of the relations between the objects under consideration. Depending on a particular application—we consider *Google problem*, *brand loyalty*, and *social status*— those interrelations give rise to transition probabilities and, hence, to the definition of a ranking as the leading *eigenvector* of a corresponding stochastic matrix. In this chapter we explain the mathematics behind ranking. First, we focus on the existence of a ranking by using the duality of linear programming. This leads to *Perron-Frobenius theorem* from linear algebra. Second, a dynamic procedure known as *PageRank* is studied. The latter enables us to approximate rankings by iterative schemes in a fast and computationally cheap manner, which is crucial for big data applications.

1.1   Motivation: Google Problem

Searching the World Wide Web is not only among the most popular, but probably also one of the oldest big data applications. Although search engines, such as e.g. the World Wide Web Worm by Mcbryan (1994), existed since early nineties, it took until Google to revolutionize the web search. The outstanding success of Google is mainly due to the idea of ranking based on the network of web pages with associated hyperlinks. The connection

© Springer-Verlag GmbH Germany, part of Springer Nature 2021
V. Shikhman, D. Müller, *Mathematical Foundations of Big Data Analytics*,
https://doi.org/10.1007/978-3-662-62521-7_1

Fig. 1.1 Network N1

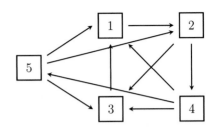

to big data is straightforward, as even the prototype dealt with a number of more than 24 million web pages and 514 hyperlinks. In what follows, we present the mathematical formulation of the Google problem, see Brin and Page (1998).

In order to provide a basic understanding, we introduce a fictitious network N1 of web pages with hyperlinks given in Fig. 1.1. Here, an arrow indicates an outgoing hyperlink from one page to another. Obviously, any reasonable order of the web pages should capture their popularity and sort them accordingly. Which pages are popular? Since we assume that the content of web pages is inaccessible, a proxy for the popularity could be derived from the number of incoming hyperlinks. At first rather naive attempt, we may rank the web pages by comparing these numbers. For the network N1 we would have the following result:

Web page:	1	2	3	4	5
Number of incoming hyperlinks:	3	2	3	1	1
Order:	I-II	III	I-II	IV-V	IV-V

The web pages 1 and 3 are the most popular, since they have three incoming hyperlinks. Let us compare the popularity of the web pages referring to 1 and 3 respectively. The web pages 3, 4, and 5 refer to 1, and the web pages 2, 4, and 5 refer to 3. Note that the web page 3 is more popular than 2 according to the number of incoming hyperlinks. Therefore, the page 1 should be declared as more popular than 3 by taking into account the second level of comparison. Indeed, together with the coinciding web pages 4 and 5 the more popular page 3 refers to 1 than the less popular web page 2 refers to 3. This consideration suggests to revise the naive approach on the number of incoming hyperlinks as follows:

A web page is the popular the more hyperlinks from *popular* web pages refer to it.

The mathematical description of this recursive principle leads to the Google problem. For that, we associate with the network consisting of n web pages the *transition* $(n \times n)$-*matrix* $P = (p_{ij})$. Its entry p_{ij} in the i-th row and j-th column expresses the probability of transition from the web page j to the web page i. In case there are no hyperlinks from

j to i, this probability vanishes, otherwise it holds:

$$p_{ij} = \frac{1}{\text{number of outgoing hyperlinks from } j}.$$

The transition matrix corresponding to the network N1 is then:

$$P = \begin{pmatrix} & \boxed{1}\;\boxed{2}\;\boxed{3}\;\boxed{4}\;\boxed{5} \\ \boxed{1} & 0 \quad 0 \quad 1 \quad 1/3 \; 1/3 \\ \boxed{2} & 1 \quad 0 \quad 0 \quad 0 \quad 1/3 \\ \boxed{3} & 0 \quad 1/2 \; 0 \quad 1/3 \; 1/3 \\ \boxed{4} & 0 \quad 1/2 \; 0 \quad 0 \quad 0 \\ \boxed{5} & 0 \quad 0 \quad 0 \quad 1/3 \; 0 \end{pmatrix}.$$

The i-th row of P consists of transition probabilities from an every web page to the i-th. The j-th column of P consists of transition probabilities from the j-th web page to the others. Thus, the entries of the j-th column sum up to one:

$$\sum_{i=1}^{n} p_{ij} = 1 \quad \text{for } j = 1, \ldots, n.$$

The latter can be written in matrix form as follows:

$$e^T \cdot P = e^T,$$

where $e = (1, \ldots, 1)^T$ stands for the n-dimensional vector of ones. Further, we denote by $x_i \geq 0$ the yet unknown *ranking* of the i-th web page, which represents its popularity. For the overall ranking we write $x = (x_1, \ldots, x_n)^T$ with $x \geq 0$ understood componentwise. Due to the Google principle, the ranking of the i-th web page can be expressed through the rankings of the web pages referring to it. The idea is namely to set the i-th ranking as the sum of the web pages' rankings times the corresponding transition probabilities:

$$x_i = \sum_{j=1}^{n} p_{ij} \cdot x_j \quad \text{for } i = 1, \ldots, n.$$

Equivalently, we write in matrix form:

$$x = P \cdot x.$$

Overall, the *Google problem* for a network with n web pages is to find a *ranking* $x \in \mathbb{R}^n$ satisfying

$$x = P \cdot x, \quad x \geq 0, \quad x \neq 0. \tag{$\mathcal{R}$}$$

Here, the corresponding $(n \times n)$-matrix P of transition probabilities is *stochastic*, i.e. its entries are nonnegative and its every column sums up to one:

$$P \geq 0, \quad e^T \cdot P = e^T.$$

1.2   Results

1.2.1   Perron-Frobenius Theorem

Let us first address the question on the existence of a ranking. This will be guaranteed by an application of Perron-Frobenius theorem. Instead of citing the latter, we rather derive its conclusion by using the duality of linear programming.

1.2.1.1 Eigenvector Problem

We start the discussion on rankings with the observation that they can be viewed as particular eigenvectors of P. Recall that $x \neq 0$ is called an *eigenvector* of the matrix P corresponding to the eigenvalue 1 if it holds:

$$P \cdot x = 1 \cdot x,$$

which coincides with the first condition in $(\mathcal{R})$. Does a stochastic matrix P have an eigenvector corresponding to the eigenvalue 1? For that, it suffices to show that the determinant of $P - I$ vanishes, where I denotes the identity matrix, see e.g. Lancaster (1969). Since the determinant is invariant under the addition of rows, we may add the rows $2, \ldots, n$ to the first one and use the fact that the matrix P is stochastic:

$$\det(P - I) = \begin{vmatrix} p_{11} - 1 & p_{12} & \cdots & p_{1n} \\ p_{21} & p_{22} - 1 & \cdots & p_{2n} \\ \vdots & \vdots & & \vdots \\ p_{n1} & p_{n2} & \cdots & p_{nn} - 1 \end{vmatrix} =$$

$$= \begin{vmatrix} \sum_{i=1}^{n} p_{i1} - 1 & \sum_{i=1}^{n} p_{i2} - 1 & \cdots & \sum_{i=1}^{n} p_{in} - 1 \\ p_{21} & p_{22} - 1 & \cdots & p_{2n} \\ \vdots & \vdots & & \vdots \\ p_{n1} & p_{n2} & \cdots & p_{nn} - 1 \end{vmatrix} = \begin{vmatrix} 0 & 0 & \cdots & 0 \\ p_{21} & p_{22} - 1 & \cdots & p_{2n} \\ \vdots & \vdots & & \vdots \\ p_{n1} & p_{n2} & \cdots & p_{nn} - 1 \end{vmatrix} = 0.$$

Hence, we are sure that any stochastic matrix possesses an eigenvector corresponding to the eigenvalue 1. Is there at least one of them with nonnegative components, i.e. satisfying the second condition in $(\mathcal{R})$? Let us examine this issue for the transition matrix P derived from the network N1. In order to compute the eigenvectors of P corresponding to the eigenvalue 1, we need to solve the above system of linear equations:

$$(P - I) \cdot x = 0,$$

where

$$P - I = \begin{pmatrix} 0 & 0 & 1 & 1/3 & 1/3 \\ 1 & 0 & 0 & 0 & 1/3 \\ 0 & 1/2 & 0 & 1/3 & 1/3 \\ 0 & 1/2 & 0 & 0 & 0 \\ 0 & 0 & 0 & 1/3 & 0 \end{pmatrix} - \begin{pmatrix} 1 & 0 & 0 & 0 & 0 \\ 0 & 1 & 0 & 0 & 0 \\ 0 & 0 & 1 & 0 & 0 \\ 0 & 0 & 0 & 1 & 0 \\ 0 & 0 & 0 & 0 & 1 \end{pmatrix} = \begin{pmatrix} -1 & 0 & 1 & 1/3 & 1/3 \\ 1 & -1 & 0 & 0 & 1/3 \\ 0 & 1/2 & -1 & 1/3 & 1/3 \\ 0 & 1/2 & 0 & -1 & 0 \\ 0 & 0 & 0 & 1/3 & -1 \end{pmatrix}.$$

The Gaussian elimination easily provides its solutions, see Exercise 1.1. The eigenvectors differ up to the multiplicative factor $t \in \mathbb{R}$:

$$x = t \cdot \begin{pmatrix} 17/3 \\ 18/3 \\ 13/3 \\ 9/3 \\ 1 \end{pmatrix}.$$

Since all of them result in the same order of the web pages, the parameter t can be chosen arbitrarily. For the sake of consistency, we normalize x by setting $t = 3/60$, so that its components sum up to one:

$$x = \frac{3}{60} \cdot \begin{pmatrix} 17/3 \\ 18/3 \\ 13/3 \\ 9/3 \\ 1 \end{pmatrix} = \begin{pmatrix} 17/60 \\ 18/60 \\ 13/60 \\ 9/60 \\ 3/60 \end{pmatrix}.$$

Note that this eigenvector of P has nonnegative components and, hence, satisfies also the second condition in $(\mathcal{R})$. The ranking x provides the following order of the web pages:

Web page:	1	2	3	4	5
Ranking:	17/60	18/60	13/60	9/60	3/60
Order:	II	I	III	IV	V

Although the web page $\boxed{2}$ has only two incoming hyperlinks, it turns out to be the most popular according to the Google principle. Confer that according to the naive approach—based on the number of incoming links—the popularity of the web page $\boxed{2}$ was inferior in comparison to the web pages $\boxed{1}$ and $\boxed{3}$.

1.2.1.2 Feasibility System

The existence of a ranking is equivalent to the feasibility of the following system:

$$x = P \cdot x, \quad x \geq 0, \quad e^T \cdot x = 1. \tag{$\mathcal{X}$}$$

Indeed, if the components of a solution of $(\mathcal{X})$ sum up to one, i.e. $e^T \cdot x = 1$, they cannot vanish all together, i.e. $x \neq 0$, hence, we have a ranking according to $(\mathcal{R})$. For the reverse implication one normalizes a ranking from $(\mathcal{R})$—by dividing its every component by their non-vanishing sum—to obtain a solution of $(\mathcal{X})$. In what follows, we show that the system $(\mathcal{X})$ is feasible. For that, we introduce in the first step its relaxed version:

$$z \geq P \cdot z, \quad z \geq 0, \quad e^T \cdot z \geq 1. \tag{$\mathcal{Z}$}$$

If the relaxation $(\mathcal{Z})$ is feasible, then so is the system $(\mathcal{X})$. To see this, let us start with a solution z of $(\mathcal{Z})$ and examine if a solution of $(\mathcal{X})$ can be defined by

$$x = \frac{z}{e^T \cdot z}.$$

For the latter it holds namely:

$$x - P \cdot x = \frac{z}{e^T \cdot z} - \frac{P \cdot z}{e^T \cdot z} = \underbrace{\frac{1}{e^T \cdot z}}_{>0} \cdot \Big(\underbrace{z - P \cdot z}_{\geq 0} \Big) \geq 0.$$

Furthermore, the sum of nonnegative components of $x - P \cdot x$ vanishes due to $e^T \cdot P = e^T$:

$$e^T \cdot (x - P \cdot x) = e^T \cdot x - e^T \cdot P \cdot x = e^T \cdot x - e^T \cdot x = 0.$$

Hence, $x - P \cdot x$ is the zero-vector and the first condition in $(\mathcal{X})$ is satisfied, i.e. $x = P \cdot x$. The other conditions in $(\mathcal{X})$ are also satisfied by x:

$$x = \frac{z}{e^T \cdot z} \geq 0, \quad e^T \cdot x = e^T \cdot \frac{z}{e^T \cdot z} = 1.$$

1.2.1.3 Linear Duality

In the second step we show that the relaxation ($\mathcal{Z}$) is feasible. This will be done by applying the duality theory of linear programming. We introduce as ($\mathcal{P}$) and ($\mathcal{D}$) the so-called primal and dual linear programming problems, respectively:

$$\min_{u \in \mathbb{R}^k} \quad c^T \cdot u \quad \text{s.t.} \quad A \cdot u \geq b, \quad u \geq 0, \tag{$\mathcal{P}$}$$

$$\max_{v \in \mathbb{R}^m} \quad b^T \cdot v \quad \text{s.t.} \quad A^T \cdot v \leq c, \quad v \geq 0. \tag{$\mathcal{D}$}$$

Here, we have the data stored in

$$c \in \mathbb{R}^k, \quad A \in \mathbb{R}^{m \times k}, \quad b \in \mathbb{R}^m.$$

If u is feasible for ($\mathcal{P}$) and v is feasible for ($\mathcal{D}$) we obtain:

$$\underbrace{c^T}_{\geq A^T \cdot v} \cdot \underbrace{u}_{\geq 0} \geq \left(A^T \cdot v \right)^T \cdot u = v^T \cdot \left(A^T \right)^T \cdot u = \underbrace{v^T}_{\geq 0} \cdot \underbrace{A \cdot u}_{\geq b} \geq v^T \cdot b = b^T \cdot v.$$

Minimizing the left hand-side over all feasible u's und maximizing the right hand-side over all feasible v's, we conclude:

$$\min \left\{ c^T \cdot u \mid A \cdot u \geq b, u \geq 0 \right\} \geq \max \left\{ b^T \cdot v \mid A^T \cdot v \leq c, v \geq 0 \right\}.$$

This inequality is known as the *weak duality* of linear programming. Weak duality means that the optimal value of the primal problem ($\mathcal{P}$) does not fall below the optimal value of the dual problem ($\mathcal{D}$). Actually, much more can be said about the optimal values of ($\mathcal{P}$) and ($\mathcal{D}$). The *strong duality* of linear programming says, see e.g. Jongen et al. (2004):

- ($\mathcal{P}$) is solvable if and only if ($\mathcal{D}$) is solvable.
- In this case the optimal values of ($\mathcal{P}$) and ($\mathcal{D}$) coincide.

1.2.1.4 Existence

In order to apply the strong duality in our context, we test the feasibility of the relaxation ($\mathcal{Z}$) by introducing the associated linear programming problem:

$$\min_{z} \quad 0^T \cdot z \quad \text{s.t.} \quad z \geq P \cdot z, \quad z \geq 0, \quad e^T \cdot z \geq 1. \tag{$\mathcal{P}_z$}$$

Note that ($\mathcal{Z}$) is feasible if and only if ($\mathcal{P}_z$) is solvable. Let us write ($\mathcal{P}_z$) in the primal form using the identity matrix I:

$$\min_{z} \quad 0^T \cdot z \quad \text{s. t.} \quad \begin{pmatrix} I - P \\ e^T \end{pmatrix} \cdot z \ge \begin{pmatrix} 0 \\ 1 \end{pmatrix}, \quad z \ge 0.$$

The strong duality of linear programming says that the primal problem $(\mathcal{P}_z)$ is solvable if and only if the corresponding dual problem is solvable:

$$\max_{y, y_{n+1}} \quad \begin{pmatrix} 0 \\ 1 \end{pmatrix}^T \cdot \begin{pmatrix} y \\ y_{n+1} \end{pmatrix} \quad \text{s. t.} \quad \begin{pmatrix} I - P \\ e^T \end{pmatrix}^T \cdot \begin{pmatrix} y \\ y_{n+1} \end{pmatrix} \le 0, \quad y, y_{n+1} \ge 0.$$

After simplifications we get for the latter:

$$\max_{y, y_{n+1}} \quad y_{n+1} \quad \text{s. t.} \quad y \le P^T \cdot y - y_{n+1} \cdot e, \quad y, y_{n+1} \ge 0. \tag{$\mathcal{D}_z$}$$

A solution of $(\mathcal{D}_z)$ can be given explicitly by

$$\bar{y} = e, \quad \bar{y}_{n+1} = 0.$$

To be sure of this, we have to test first if it satisfies the constraint in $(\mathcal{D}_z)$:

$$e = \bar{y} \le P^T \cdot \bar{y} - \bar{y}_{n+1} \cdot e = P^T \cdot e - 0 \cdot e = e.$$

Additionally, we have to show that the optimal value of $(\mathcal{D}_z)$ cannot exceed $\bar{y}_{n+1} = 0$. For that, let us consider the componentwise maximum of the constraint in $(\mathcal{D}_z)$:

$$\max_{1 \le i \le n} y_i \le \max_{1 \le i \le n} \left(P^T \cdot y \right)_i - y_{n+1} \le \max_{1 \le i \le n} \underbrace{\left(P^T \cdot e \right)_i}_{=1} \cdot \max_{1 \le i \le n} y_i - y_{n+1} = \max_{1 \le j \le n} y_i - y_{n+1}.$$

It follows:

$$y_{n+1} \le 0.$$

We conclude that the dual problem $(\mathcal{D}_z)$ is solvable, hence, due to the strong duality so is the primal problem $(\mathcal{P}_z)$. This provides the feasibility of the relaxation $(\mathcal{Z})$, which implies the feasibility of the system $(\mathcal{X})$. Thus, we have shown that a stochastic matrix always has an eigenvector with nonnegative components corresponding to the eigenvalue 1. The latter result is a special case of *Perron-Frobenius theorem*, see also Lancaster (1969).

1.2.2 PageRank

After ensuring rankings' existence, we introduce and analyze iteration schemes aiming to their efficient computation. For that, let us model the users' surfing behavior on the network of web pages.

1.2.2.1 Web Surfing

We denote by $x_i(t) \geq 0$ the share of users visiting the i-th web page at time t. For the overall *distribution* of users we write

$$x(t) = (x_1(t), \ldots, x_n(t))^T,$$

such that it satisfies

$$x(t) \geq 0, \quad e^T \cdot x(t) = 1.$$

The share $x_i(t+1)$ of users visiting at time $t+1$ the i-th web page can be calculated iteratively. By summing up all the shares $x_j(t)$ of users visiting at time t those web pages j, which refer to the i-th with the corresponding transition probability p_{ij}, we obtain:

$$x_i(t+1) = \sum_{j=1}^{n} p_{ij} \cdot x_j(t), \quad i = 1, \ldots, n.$$

Written in matrix form, we get the following iteration scheme:

$$x(t+1) = P \cdot x(t) \quad \text{for } t = 1, 2, \ldots. \tag{$\mathcal{I}$}$$

This dynamics is well-defined if we start with a given distribution $x(1)$. Then, $x(t+1)$ becomes also a distribution. This can be shown by induction. Since the matrix P is stochastic, we have:

$$x(t+1) = \underbrace{P}_{\geq 0} \cdot \underbrace{x(t)}_{\geq 0} \geq 0,$$

and

$$e^T \cdot x(t+1) = \underbrace{e^T \cdot P}_{=e^T} \cdot x(t) = e^T \cdot x(t) = 1.$$

Let us assume for a moment that the iteration scheme $(\mathcal{I})$ converges, i.e.

$$x(t+1) \to x \quad \text{for } t \to \infty.$$

Taking the limit in

$$x(t+1) = P \cdot x(t), \quad x(t+1) \geq 0, \quad e^T \cdot x(t+1) = 1,$$

we conclude that x solves the system $(\mathcal{X})$:

$$x = P \cdot x, \quad x \geq 0, \quad e^T \cdot x = 1.$$

Hence, the iteration scheme $(\mathcal{I})$—if convergent—leads to a ranking. This means that the long-term surfing on the network eventually results in a ranking. In line with this, a ranking gives the web pages' *stationary* shares of users, which do not change under further surfing anymore.

1.2.2.2 Oscillation

Unfortunately, the iteration scheme $(\mathcal{I})$ does not need to converge in general, on the contrary, the oscillation may occur. This is e.g. the case for the network N1. We illustrate the oscillating phenomenon on a much simpler network N2 given in Fig. 1.2. The transition matrix and the corresponding ranking for the network N2 are

$$P = \begin{pmatrix} & \boxed{1}\ \boxed{2}\ \boxed{3} \\ \boxed{1} & 0 \quad 1 \quad 1 \\ \boxed{2} & 1/2 \quad 0 \quad 0 \\ \boxed{3} & 1/2 \quad 0 \quad 0 \end{pmatrix}, \quad x = \begin{pmatrix} 1/2 \\ 1/4 \\ 1/4 \end{pmatrix}.$$

Nevertheless, the iteration scheme $(\mathcal{I})$ oscillates. To see this, we compute its iterates starting from an arbitrary distribution with $a, b, c \geq 0$ and $a + b + c = 1$:

$$x(1) = \begin{pmatrix} a \\ b \\ c \end{pmatrix}, \quad x(2) = \begin{pmatrix} b+c \\ a/2 \\ a/2 \end{pmatrix}, \quad x(3) = \begin{pmatrix} a \\ (b+c)/2 \\ (b+c)/2 \end{pmatrix}, \quad x(4) = \begin{pmatrix} b+c \\ a/2 \\ a/2 \end{pmatrix} \quad \text{etc.}$$

Fig. 1.2 Network N2

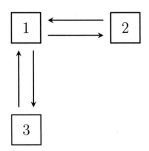

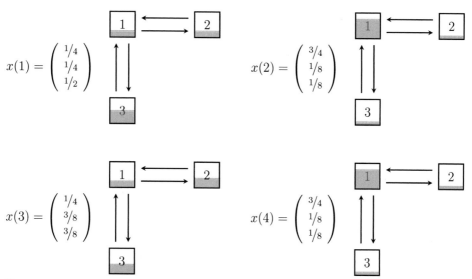

Fig. 1.3 Oscillation in network N2

We visualize this oscillation phenomenon for the starting distribution with $a = b = 1/4$ and $c = 1/2$ in Fig. 1.3.

1.2.2.3 Regularization

In order to avoid oscillation, it is convenient to regularize the stochastic matrix P by

$$P_\alpha = (1 - \alpha) \cdot P + \alpha \cdot E,$$

where $\alpha \in (0, 1)$ is a sufficiently small parameter, and E is the stochastic $(n \times n)$-matrix with the entries equal to $1/n$, i.e.

$$E = \frac{1}{n} \cdot e \cdot e^T.$$

Note that P_α remains stochastic as the convex combination of stochastic matrices:

$$P_\alpha = (1 - \alpha) \cdot \underbrace{P}_{\geq 0} + \alpha \cdot \underbrace{E}_{\geq 0} \geq 0,$$

$$e^T \cdot P_\alpha = (1 - \alpha) \cdot \underbrace{e^T \cdot P}_{=e} + \alpha \cdot \underbrace{e^T \cdot E}_{=e} = (1 - \alpha) \cdot e + \alpha \cdot e = e.$$

The interpretation of P_α in terms of surfing is straightforward. With probability $1 - \alpha$ the users follow the hyperlinks of the network with transition probabilities given in P. But sometimes—with probability α—they start surfing anew according to the uniform transition probabilities given in E. This leads to the iteration scheme known as *PageRank*, see Brin and Page (1998):

$$x_\alpha(t+1) = P_\alpha \cdot x_\alpha(t) \quad \text{for } t = 1, 2, \ldots . \tag{$\mathcal{I}_\alpha$}$$

1.2.2.4 Convergence Analysis of $(\mathcal{I}_\alpha)$

We show that the iteration scheme $(\mathcal{I}_\alpha)$ converges, moreover, its limit is the so-called Google ranking. A vector $x_\alpha \in \mathbb{R}^n$ is called *Google ranking* if it solves the regularized system:

$$x_\alpha = P_\alpha \cdot x_\alpha, \quad x_\alpha \geq 0, \quad e^T \cdot x_\alpha = 1. \tag{$\mathcal{X}_\alpha$}$$

Since the matrix P_α is stochastic, the regularized system $(\mathcal{X}_\alpha)$ is feasible, i.e. a Google ranking x_α always exists. In what follows, we make use of the *Manhattan norm* in $\mathbb{R}^n$:

$$\|x\|_1 = \sum_{i=1}^n |x_i| = e^T \cdot |x|.$$

Using the triangle inequality, we estimate the distance between $x_\alpha(t+1)$ and x_α with respect to the Manhattan norm:

$$\|x_\alpha(t+1) - x_\alpha\|_1 = \|P_\alpha \cdot x_\alpha(t) - P_\alpha \cdot x_\alpha\|_1 = \|P_\alpha \cdot (x_\alpha(t) - x_\alpha)\|_1$$

$$= \|((1-\alpha) \cdot P + \alpha \cdot E) \cdot (x_\alpha(t) - x_\alpha)\|_1$$

$$= \left\| (1-\alpha) \cdot P \cdot (x_\alpha(t) - x_\alpha) + \frac{\alpha}{n} \cdot \left(e \cdot \underbrace{e^T \cdot x_\alpha(t)}_{=1} - e \cdot \underbrace{e^T \cdot x_\alpha}_{=1} \right) \right\|_1$$

$$= (1-\alpha) \cdot \|P \cdot (x_\alpha(t) - x_\alpha)\|_1 = (1-\alpha) \cdot e^T \cdot \underbrace{|P \cdot (x_\alpha(t) - x_\alpha)|}_{\leq P \cdot |x_\alpha(t) - x_\alpha|}$$

$$\leq (1-\alpha) \cdot \underbrace{e^T \cdot P}_{=e^T} \cdot |x_\alpha(t) - x_\alpha| = (1-\alpha) \cdot \|x_\alpha(t) - x_\alpha\|_1 .$$

Applying this inequality successively, we obtain:

$$\|x_\alpha(t+1) - x_\alpha\|_1 \leq (1-\alpha) \cdot \|x_\alpha(t) - x_\alpha\|_1 \leq \ldots \leq (1-\alpha)^t \cdot \|x(1) - x_\alpha\|_1 .$$

Due to $\alpha \in (0, 1)$, it finally holds:

$$\|x_\alpha(t+1) - x_\alpha\|_1 \leq (1-\alpha)^t \cdot \|x(1) - x_\alpha\|_1 \to 0 \quad \text{for } t \to \infty.$$

From here we conclude that the iteration scheme $(\mathcal{I}_\alpha)$ converges towards the unique Google ranking x_α. Moreover, the convergence is *linear* with the rate $1 - \alpha$. This means that every iteration in $(\mathcal{I}_\alpha)$ approaches the Google ranking at least by the factor $1 - \alpha$. In Brin and Page (1998) it is suggested to choose the parameter α not too large, so that users follow the underlying network with a sufficiently high probability $1 - \alpha$. Usually, $\alpha \approx 0.15$ is taken. Then, a reasonable convergence rate $1 - \alpha \approx 0.85$ for the iteration scheme $(\mathcal{I}_\alpha)$ is guaranteed.

1.2.2.5 Google Ranking

Let us compute the Google ranking for the network N1. By choosing $\alpha = 0.15$, the regularized matrix is

$$P_{0.15} = (1 - 0.15) \cdot \begin{pmatrix} 0 & 0 & 1 & 1/3 & 1/3 \\ 1 & 0 & 0 & 0 & 1/3 \\ 0 & 1/2 & 0 & 1/3 & 1/3 \\ 0 & 1/2 & 0 & 0 & 0 \\ 0 & 0 & 0 & 1/3 & 0 \end{pmatrix} + 0.15 \cdot \begin{pmatrix} 1/5 & 1/5 & 1/5 & 1/5 & 1/5 \\ 1/5 & 1/5 & 1/5 & 1/5 & 1/5 \\ 1/5 & 1/5 & 1/5 & 1/5 & 1/5 \\ 1/5 & 1/5 & 1/5 & 1/5 & 1/5 \\ 1/5 & 1/5 & 1/5 & 1/5 & 1/5 \end{pmatrix} .$$

By calculating the latter, we get:

$$P_{0.15} = \begin{pmatrix} 18/600 & 18/600 & 528/600 & 188/600 & 188/600 \\ 528/600 & 18/600 & 18/600 & 18/600 & 188/600 \\ 18/600 & 273/600 & 18/600 & 188/600 & 188/600 \\ 18/600 & 273/600 & 18/600 & 18/600 & 18/600 \\ 18/600 & 18/600 & 18/600 & 188/600 & 18/600 \end{pmatrix} .$$

Its normalized eigenvector corresponding to the eigenvalue 1 is

$$x_{0.15} \approx \begin{pmatrix} 0.276 \\ 0.285 \\ 0.215 \\ 0.151 \\ 0.073 \end{pmatrix} .$$

The Google ranking $x_{0.15}$ provides the same order of the web pages as before:

Web page:	1	2	3	4	5
Google ranking:	0.276	0.285	0.215	0.151	0.073
Order:	II	I	III	IV	V

Moreover, the Google ranking $x_{0.15}$ is realized by users who surf on the network according to the iteration scheme $(\mathcal{X}_{0.15})$. With probability 0.85 they stay in the network and continue surfing. From time to time—with probability 0.15—they leave the network and enter a new search. On a long-term basis, the share of the most popular web page $\boxed{2}$ will then amount to approximately 28.5 % of users, while the share of the least popular web page $\boxed{5}$ becomes 7.3 % etc.

1.2.2.6 Global Authority

The convergence analysis of PageRank motivates to state a sufficient condition under which the iteration scheme $(\mathcal{I})$ successively approaches a ranking. For that, we assume that the transition matrix P has at least one row with all entries to be positive, see Nesterov and Nemirovski (2015). This means that there exists a web page, called *global authority*, which can be reached with non-zero probability from all the others within the network. Let the i-th web page be such a global authority. Then, it holds:

$$p_{ij} > 0 \quad \text{for all } j = 1, \dots, n.$$

We consider the smallest entry of the i-th row:

$$\widetilde{\alpha} = \min_{j=1,\dots,n} p_{ij} > 0.$$

Let us denote the stochastic $(n \times n)$-matrix with ones at its i-th row and zeros elsewhere by

$$E_i = e_i \cdot e_i^T,$$

where e_i is the i-th coordinate vector of $\mathbb{R}^n$. We try to represent P as a convex combination of stochastic matrices:

$$P = (1 - \widetilde{\alpha}) \cdot \widetilde{P} + \widetilde{\alpha} \cdot E_i$$

by setting

$$\widetilde{P} = \frac{1}{1 - \widetilde{\alpha}} \cdot (P - \widetilde{\alpha} \cdot E_i) \, .$$

It is straightforward to see that $\widetilde{P}$ is stochastic:

$$\widetilde{p}_{ij} = \frac{1}{1 - \widetilde{\alpha}} \cdot (p_{ij} - \widetilde{\alpha}) = \frac{1}{1 - \widetilde{\alpha}} \cdot \left(p_{ij} - \min_{j=1,\dots,n} p_{ij} \right) \geq 0,$$

$$\widetilde{p}_{kj} = \frac{1}{1 - \widetilde{\alpha}} \cdot p_{kj} \geq 0 \quad \text{for } k \neq i,$$

and

$$e^T \cdot \tilde{P} = \frac{1}{1 - \tilde{\alpha}} \cdot \left(e^T \cdot P - \tilde{\alpha} \cdot e^T \cdot E_i \right) = \frac{1}{1 - \tilde{\alpha}} \cdot \left(e^T - \tilde{\alpha} \cdot e^T \right) = e^T.$$

Analogously to PageRank, we estimate for any ranking x:

$$\|x(t+1) - x\|_1 = \|P \cdot (x(t) - x)\|_1 = \left\| \left((1 - \tilde{\alpha}) \cdot \tilde{P} + \tilde{\alpha} \cdot E_i \right) \cdot (x(t) - x) \right\|_1$$

$$= \left\| (1 - \tilde{\alpha}) \cdot \tilde{P} \cdot (x(t) - x) + \tilde{\alpha} \cdot \underbrace{(E_i \cdot x(t) - E_i \cdot x)}_{= e_i - e_i = 0} \right\|_1$$

$$= (1 - \tilde{\alpha}) \cdot \left\| \tilde{P} \cdot (x(t) - x) \right\|_1 = (1 - \tilde{\alpha}) \cdot e^T \cdot \underbrace{\left| \tilde{P} \cdot (x(t) - x) \right|}_{\leq \tilde{P} \cdot |x(t) - x|}$$

$$\leq (1 - \tilde{\alpha}) \cdot \underbrace{e^T \cdot \tilde{P}}_{= e^T} \cdot |x(t) - x| = (1 - \tilde{\alpha}) \cdot \|x(t) - x\|_1.$$

Due to $0 < \tilde{\alpha} \leq 1$, we recursively have:

$$\|x(t+1) - x\|_1 \leq (1 - \tilde{\alpha})^t \cdot \|x(1) - x\|_1 \to 0 \quad \text{for } t \to \infty.$$

Hence, the convergence rate of $(\mathcal{I})$ is $1 - \tilde{\alpha}$, and its limit x is the unique ranking.

1.3   Case Study: Brand Loyalty

Let us model the brand loyalty by means of rankings. To become specific, we look at a shoe market where three companies—Adidas, Nike, and Puma—fight for costumers. For the sake of simplicity, we assume that each company is associated with only one brand, labeled by A, N, and P. In every period of time customers buy exactly one pair of shoes by choosing one of the brands. The data on their shopping behavior is given as follows:

Consumer 1	A A A A A A N A P A A A
Consumer 2	P N N N N N N N N A
Consumer 3	P P P P P P P N P A A

Fig. 1.4 Weighted network
N3

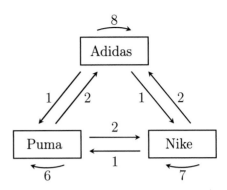

Such data sets could be stored by big online shops like Amazon. From these purchases it is possible to extract some valuable information on the market structure. In particular, companies may be interested in learning about their future shares of customers.

Task 1 Based on the shopping behavior from above, construct a weighted network N3 of customers' transitions between the brands. For that, the weight of a link is associated with the number of transitions from one brand to another performed by customers.

Hint 1 For the weighted network N3 see Fig. 1.4.

Task 2 Based on the weighted network N3, construct a stochastic matrix $P = (p_{ij})$ of transition probabilities. Its general entry p_{ij} corresponds to the number of transitions from the j-th to the i-th brand, which is set in relation to the overall transitions from the former.

Hint 2 The transition matrix is

$$P = \begin{pmatrix} & \text{Adidas} & \text{Nike} & \text{Puma} \\ \text{Adidas} & 8/10 & 2/10 & 2/10 \\ \text{Nike} & 1/10 & 7/10 & 2/10 \\ \text{Puma} & 1/10 & 1/10 & 6/10 \end{pmatrix}.$$

Task 3 Denote by $x(t) = (x_1(t), x_2(t), x_3(t))^T$ the companies' shares of customers at time t. Formulate the shares' update $(\mathcal{I})$ by means of the transition matrix P.

Hint 3 The shares of customers are updated by

$$x(t+1) = P \cdot x(t) \quad \text{for } t = 1, 2, \ldots,$$

or, equivalently:

$$
\begin{pmatrix} x_1(t+1) \\ x_2(t+1) \\ x_3(t+1) \end{pmatrix} = \begin{pmatrix} 8/10 & 2/10 & 2/10 \\ 1/10 & 7/10 & 2/10 \\ 1/10 & 1/10 & 6/10 \end{pmatrix} \cdot \begin{pmatrix} x_1(t) \\ x_2(t) \\ x_3(t) \end{pmatrix} \quad \text{for } t = 1, 2, \ldots.
$$

Here, $x(1)$ is a starting distribution of customers, i.e.

$$
x(1) \geq 0, \quad e^T \cdot x(1) = 1.
$$

Task 4 By using the global authority condition, show that the update $(\mathcal{I})$ converges. Provide the convergence rate, find its limit and interpret the latter as a ranking.

Hint 4 Since all rows of the transition matrix P are positive, every brand is a global authority. By choosing for the latter e.g. Adidas, the smallest entry of the first row is $\tilde{\alpha} = 2/10$. We therefore conclude that the convergence rate of $(\mathcal{I})$ is at least

$$
1 - \tilde{\alpha} = 1 - \frac{2}{10} = \frac{8}{10}.
$$

The limit of $(\mathcal{I})$ can be computed as the eigenvector of P corresponding to the eigenvalue 1:

$$
x = \begin{pmatrix} 1/2 \\ 3/10 \\ 1/5 \end{pmatrix}.
$$

The ranking x gives the companies' shares of customers, provided the underlying consumption behaviour is due to the transition probabilities given in P. On a long-term basis, Adidas accumulates 50 %, Nike 30 %, and Puma 20 % of customers.

Task 5 Write P as a convex combination of a suitable stochastic matrix $\bar{P}$ and E, so that with $\bar{\alpha} \in (0, 1]$ it holds:

$$
P = (1 - \bar{\alpha}) \cdot \bar{P} + \bar{\alpha} \cdot E.
$$

Improve the theoretically guaranteed convergence rate from Task 4 by using this representation.

Hint 5 Since the transition matrix is positive, Exercise 1.4 may be applied to deliver:

$$\begin{pmatrix} 8/10 & 2/10 & 2/10 \\ 1/10 & 7/10 & 2/10 \\ 1/10 & 1/10 & 6/10 \end{pmatrix} = \left(1 - \frac{3}{10}\right) \cdot \begin{pmatrix} 1 & 1/7 & 1/7 \\ 0 & 6/7 & 1/7 \\ 0 & 0 & 5/7 \end{pmatrix} + \frac{3}{10} \cdot \begin{pmatrix} 1/3 & 1/3 & 1/3 \\ 1/3 & 1/3 & 1/3 \\ 1/3 & 1/3 & 1/3 \end{pmatrix}.$$

According to Exercise 1.4, the convergence rate becomes now

$$1 - \bar{\alpha} = 1 - \frac{3}{10} = \frac{7}{10} < \frac{8}{10} = 1 - \frac{2}{10} = 1 - \tilde{\alpha}.$$

1.4   Exercises

Exercise 1.1 (Rankings) Compute the (Google) ranking for the network N1.

Exercise 1.2 (Cesàro Mean) Let a stochastic matrix P be given. The sequence of *Cesàro means* is defined by

$$\bar{x}(s) = \frac{1}{s} \cdot \sum_{t=1}^{s} x(t),$$

where $x(1), \ldots, x(s)$ are formed according to $(\mathcal{I})$ with a starting distribution $x(1)$, i.e.

$$x(t+1) = P \cdot x(t) \quad \text{for } t = 1, 2, \ldots.$$

(i) Show that the Cesàro mean $\bar{x}(s)$ is a distribution, i.e.

$$\bar{x}(s) \geq 0, \quad e^T \cdot \bar{x}(s) = 1 \quad \text{for } s = 1, 2, \ldots.$$

(ii) Establish the inequality

$$\|\bar{x}(s) - P \cdot \bar{x}(s)\|_1 \leq \frac{2}{s},$$

providing that the sequence of Cesàro means approaches the set of rankings.

Exercise 1.3 (Permutation Matrices) Let a network with n web pages be given. The only hyperlinks lead from the web page i to the web page $i + 1$ for all $i = 1, \ldots, n - 1$, and from the web page n to the web page 1. Perform the analysis of this network along the following lines:

(i) Construct the transition matrix.
(ii) Find the unique ranking.
(iii) Show that the iteration scheme ($\mathcal{I}$) oscillates.
(iv) Show that the sequence of Cesàro means converges.

Exercise 1.4 (Positive Matrices) Within a network, let the transition probability between any two pages be positive, i.e. for the transition matrix it holds $P > 0$. Perform the analysis of such a network along the following lines:

(i) For the smallest entry of P show:

$$\min_{1 \le i, j \le n} p_{ij} \le \frac{1}{n}.$$

(ii) Provide the representation:

$$P = (1 - \bar{\alpha}) \cdot \bar{P} + \bar{\alpha} \cdot E,$$

where the matrix $\bar{P}$ is stochastic, and

$$\bar{\alpha} = n \cdot \min_{1 \le i, j \le n} p_{ij}.$$

(iii) For the iterations from ($\mathcal{I}$) and a ranking x estimate:

$$\|x(t + 1) - x\|_1 \le (1 - \bar{\alpha}) \cdot \|x(t) - x\|_1.$$

(iv) Derive the convergence rate for the iteration scheme ($\mathcal{I}$) towards the unique ranking.

Exercise 1.5 (Social Status) The following Facebook data became available:

- friend $\boxed{1}$ receives 2 likes from friend $\boxed{3}$ and 1 like from friend $\boxed{4}$,
- friend $\boxed{2}$ receives 5 likes from friend $\boxed{1}$, 2 likes from friend $\boxed{3}$, and 1 like from friend $\boxed{4}$,
- friend $\boxed{3}$ receives 7 likes from friend $\boxed{2}$ and 1 like from friend $\boxed{4}$,
- friend $\boxed{4}$ receives 4 likes from friend $\boxed{1}$, 2 likes from friend $\boxed{2}$, and 4 like from friend $\boxed{3}$.

Who has the highest social status within this Facebook group?

Exercise 1.6 (Exchange Economy, See Gale (1960)) Within an exchange economy the producer P_i manufactures exactly one entity of the good G_i, $i = 1, \ldots, n$. For that, the quantities $a_{ij} \geq 0$ of the goods G_j, $j = 1, \ldots, n$, are employed. The equilibrium prices of goods are such, that the corresponding production costs do not exceed the producers' revenues. Interpret the vector of equilibrium prices as a ranking. How is it possible to adjust prices of goods in order to reach equilibrium?

Online Learning

2

In a world where automatic data collection becomes ubiquitous, we have to deal more and more often with data flow rather than with data sets. Whether we consider pricing of goods, portfolio selection, or expert advice, a common feature emerges: huge amounts of dynamic data need to be understood and quickly processed. In order to cope with this issue, the paradigm of *online learning* has been introduced. According to the latter, data becomes available in a sequential order and is used to update our decision at each iteration step. This is in strong contrast with *batch learning* which generates the best decision by learning on the entire training data set at once. For those applications, where the available amount of data truly explodes, it has become convenient to apply online learning techniques. Crucial for online learning is the question on how to measure the quality of the implemented decisions. For that, the notion of *regret*, known from decision theory, has been introduced. Loosely speaking, regret compares the losses caused by active decision strategies over time, on the one hand, with the losses caused by a passive decision strategy in hindsight, on the other hand. Surprisingly enough, online learning techniques allow to drive the average regret to zero as time progresses. In this chapter we explain the mathematics behind. First, we introduce some auxiliary notions from convex analysis, namely, those of dual norm, prox-function, and Bregman divergence. Second, we present an online learning technique called *online mirror descent*. Under convexity assumptions, an optimal rate of convergence for the corresponding regret is derived. We elaborate on the versions of the online mirror descent algorithm in entropic and Euclidean setups. In particular, the entropic setup enables us to online *portfolio selection* and to prediction with *expert advice*. The Euclidean setup leads to *online gradient descent*.

© Springer-Verlag GmbH Germany, part of Springer Nature 2021
V. Shikhman, D. Müller, *Mathematical Foundations of Big Data Analytics*,
https://doi.org/10.1007/978-3-662-62521-7_2

2.1   Motivation: Portfolio Selection

The problem of portfolio selection is fundamental in finance. Its aim is to create an optimal portfolio which reflects the preferences of an investor in terms of risk and return. A typical investor is thus trying to minimize the risk of a portfolio and to increase its expected return. Unfortunately, risk and expected return of a portfolio are coupled. Low-risk portfolios usually generate low returns. Conversely, if you want to generate a considerably high return, there is no way around, but to take a lot of risk when selecting a portfolio. Risk minimization and return maximization must therefore be balanced. The classical trade-off is due to Harry Markowitz who has been awarded the Nobel Prize for Economics for his theory of *modern portfolio selection* in 1990. Markowitz (1952) suggests to assemble a portfolio of assets such that the risk of a portfolio is minimized and a targeted level of expected return is guaranteed. It remains to mention that the variance of a portfolio serves as a proxy for risk. For expected return its mean can be taken. From the practical point of view, the mean, variance, and correlation parameters of asset prices should be then statistically estimated. One possibility is to rely upon historical data, e.g. by taking the recent history of asset prices into account. Here, we follow an alternative approach referred to as *universal portfolio selection* from Vovk and Watkins (1998). The universal portfolio selection is based on the idea that investing can be seen as a process of repeated decision making with due regard to the changes of asset prices. We proceed with a precise mathematical formulation of the latter approach.

Let assets $1, \ldots, n$ be traded on a stock market, where short selling is not possible. We consider investment strategies for the time period t. At the beginning of t-th time period an investor has to distribute the available wealth $W(t)$ among the assets by fixing their *portfolio* shares:

$$x(t) = (x_1(t), \ldots, x_n(t))^T ,$$

where

$$x_i(t) \geq 0 \quad \text{for } i = 1, \ldots, n, \quad \text{and} \quad \sum_{i=1}^{n} x_i(t) = 1.$$

Equivalently, we write $x(t) \in \Delta$ by using for short the *simplex*

$$\Delta = \left\{ x \in \mathbb{R}^n \;\middle|\; x \geq 0 \text{ and } e^T \cdot x = 1 \right\}.$$

Note that the investor will be able to evaluate the success of the portfolio $x(t)$ just at the end of the t-th time period. Then, the prices of assets will be publicly available:

$$p(t) = (p_1(t), \ldots, p_n(t))^T ,$$

where $p_i(t)$ denotes the price of the i-th asset, $i = 1, \ldots, n$. What is the investor's wealth $W(T)$ after T time periods if starting by $W(0)$? In order to compute $W(T)$, we denote the vector of assets' *returns* by

$$r(t) = (r_1(t), \ldots, r_n(t))^T ,$$

where we set

$$r_i(t) = \frac{p_i(t)}{p_i(t-1)}, \quad i = 1, \ldots, n.$$

Thus, the overall return of the portfolio $x(t)$ sums up to

$$r^T(t) \cdot x(t) = \sum_{i=1}^{n} r_i(t) \cdot x_i(t).$$

Finally, we get the investor's wealth after T time periods:

$$W(T) = W(T-1) \cdot r^T(T) \cdot x(T) = \ldots = W(0) \cdot \prod_{t=1}^{T} r^T(t) \cdot x(t).$$

The wealth $W(T)$ results from an active investment strategy without having any information on the future assets' prices. In fact, despite of not knowing the assets' returns $r(t)$, the portfolio $x(t)$ has to be realized at the beginning of the t-th time period, see Fig. 2.1.

Now, it is crucial to estimate whether the portfolios $x(t), t = 1, \ldots, T$, are efficient. For that, we construct a constant rebalancing portfolio as a benchmark. A *constant rebalancing portfolio* maximizes the investor's wealth after T time periods if the future returns are available in hindsight:

$$\max_{x \in \Delta} \ W_x(T),$$

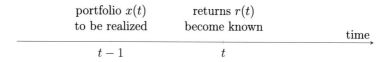

Fig. 2.1 Portfolio selection at t-th time period

where

$$W_x(T) = W(0) \cdot \prod_{t=1}^{T} r^T(t) \cdot x.$$

This maximal wealth results from a passive investment strategy in case the information on the future assets' returns is available. Let us compare the performance of passive and active investment strategies over T time periods. For that, it is convenient to take the logarithmic ratio of the corresponding wealths per period:

$$\mathcal{R}(T) = \frac{1}{T} \cdot \ln \left(\frac{\max\limits_{x \in \Delta} W_x(T)}{W(T)} \right).$$

After simplifying, we obtain:

$$\mathcal{R}(T) = \frac{1}{T} \cdot \ln \left(\max_{x \in \Delta} W_x(T) \right) - \frac{1}{T} \cdot \ln W(T) = \frac{1}{T} \cdot \max_{x \in \Delta} \ln W_x(T) - \frac{1}{T} \cdot \ln W(T)$$

$$= \max_{x \in \Delta} \frac{1}{T} \cdot \ln \left(W(0) \cdot \prod_{t=1}^{T} r^T(t) \cdot x \right) - \frac{1}{T} \cdot \ln \left(W(0) \cdot \prod_{t=1}^{T} r^T(t) \cdot x(t) \right)$$

$$= \max_{x \in \Delta} \frac{1}{T} \cdot \sum_{t=1}^{T} \ln \left(r^T(t) \cdot x \right) - \frac{1}{T} \cdot \sum_{t=1}^{T} \ln \left(r^T(t) \cdot x(t) \right)$$

$$= \frac{1}{T} \cdot \sum_{t=1}^{T} -\ln \left(r^T(t) \cdot x(t) \right) - \min_{x \in \Delta} \frac{1}{T} \cdot \sum_{t=1}^{T} -\ln \left(r^T(t) \cdot x \right).$$

This efficiency measure can be interpreted as regret in the context of online optimization:

$$\mathcal{R}(T) = \frac{1}{T} \cdot \sum_{t=1}^{T} f_t(x(t)) - \min_{x \in X} \frac{1}{T} \cdot \sum_{t=1}^{T} f_t(x),$$

where we have

$$X = \Delta, \quad f_t(x) = -\ln \left(r^T(t) \cdot x \right).$$

In online optimization, an agent successively makes feasible decisions $x \in X$. However, as a decision is made, the associated outcome remains unknown to the agent. Just after committing to a decision $x(t)$, the decision maker suffers a loss $f_t(x(t))$. These losses usually depend on the environment and cannot be estimated by the decision maker beforehand. They may be even chosen by an adversary and, in particular, depend on the actions taken by the decision maker. How does an agent know that the realized decisions $x(t)$, $t = 1, \ldots, T$, are efficient? This is done via *average regret* $\mathcal{R}(T)$ which measures the difference between the losses associated with the decisions proposed by the agent over time, and the losses generated by the optimal constant decision in hindsight. In what follows, our goal will be to adjust decisions $x(t)$, $t = 1, \ldots, T$, in such a way, so that the average regret $\mathcal{R}(T)$ becomes small, at least asymptotically as $T \to \infty$.

2.2   Results

2.2.1   Online Mirror Descent

We recall basic notions from convex analysis, see e.g. Nesterov (2018) for further details. By using convexity, a particular online learning technique, called online mirror descent, will be presented and analyzed.

2.2.1.1 Norm

A *norm* on $\mathbb{R}^n$ is a nonnegative function $\| \cdot \| : \mathbb{R}^n \to \mathbb{R}$ with the following properties ($\alpha \in \mathbb{R}$, $x, y \in \mathbb{R}^n$):

- Positive definite: $\|x\| = 0$ if and only if $x = 0$.
- Absolutely homogeneous: $\|\alpha \cdot x\| = |\alpha| \cdot \|x\|$.
- Triangle inequality: $\|x + y\| \leq \|x\| + \|y\|$.

For an arbitrary norm $\| \cdot \|$ we define the corresponding *dual norm* on $\mathbb{R}^n$:

$$\|g\|_* = \max_{\|x\| \leq 1} g^T \cdot x.$$

The dual norm inherits the properties of positive definiteness, absolute homogeneity, and triangle inequality, see Exercise 2.1. For any vectors $x \neq 0$, $g \in \mathbb{R}^n$ we have further:

$$\frac{|g^T \cdot x|}{\|x\|} = \left| g^T \cdot \frac{x}{\|x\|} \right| \leq \max_{\|x\| \leq 1} g^T \cdot x = \|g\|_*.$$

The *Hölder inequality* is thus derived:

$$\left| g^T \cdot x \right| \leq \|g\|_* \cdot \|x\|. \tag{$\mathcal{H}$}$$

2.2.1.2 Convexity

A set $X \subset \mathbb{R}^n$ is *convex* if with every two points the line segment between them is included in the set, i.e. for all $x, y \in X$ and $\lambda \in [0, 1]$ it holds:

$$\lambda \cdot x + (1 - \lambda) \cdot y \in X.$$

A continuously differentiable function $f : X \to \mathbb{R}$ is *convex* if every tangential hyperplane to its graph lies below the graph, i.e. for all $x, y \in X$ it holds:

$$f(y) \geq f(x) + \nabla^T f(x) \cdot (y - x), \tag{$\mathcal{C}^1$}$$

where the gradient of f at x is given by the vector of partial derivatives:

$$\nabla f(x) = \left(\frac{\partial f}{\partial x_1}(x), \ldots, \frac{\partial f}{\partial x_n}(x) \right)^T.$$

If f happens to be twice continuously differentiable, its convexity can be characterized by the nonnegative curvature of the graph, i.e. for all $x \in X$ and $\xi \in \mathbb{R}^n$ it holds:

$$\xi^T \cdot \nabla^2 f(x) \cdot \xi \geq 0, \tag{$\mathcal{C}^2$}$$

where the second-order derivative of f at x is given by the Hesse matrix

$$\nabla^2 f(x) = \begin{pmatrix} \frac{\partial^2 f}{\partial x_1^2}(x) & \cdots & \frac{\partial^2 f}{\partial x_1 \partial x_n}(x) \\ \vdots & & \vdots \\ \frac{\partial^2 f}{\partial x_n \partial x_1}(x) & \cdots & \frac{\partial^2 f}{\partial x_n^2}(x) \end{pmatrix}.$$

Now, we turn our attention to the optimization problem

$$\min_{x \in X} f(x),$$

where $X \subset \mathbb{R}^n$ is a convex and closed set, and $f : X \to \mathbb{R}$ is a convex and differentiable function. We state the corresponding optimality condition in variational form:

$$\bar{x} \text{ minimizes } f \text{ on } X \text{ if and only if } \nabla^T f(\bar{x}) \cdot (x - \bar{x}) \geq 0 \text{ for any } x \in X. \tag{$\mathcal{OPT}$}$$

For the sufficiency part, we use the convexity of f to deduce for any $x \in X$:

$$f(x) \geq f(\bar{x}) + \underbrace{\nabla^T f(\bar{x}) \cdot (x - \bar{x})}_{\geq 0} \geq f(\bar{x}).$$

Hence, $\bar{x}$ minimizes f on X. For the necessity part, let there exist $x \in X$ such that

$$\nabla^T f(\bar{x}) \cdot (x - \bar{x}) < 0.$$

For the convex combination of x and $\bar{x}$ with $\lambda \in [0, 1]$ it holds:

$$z(\lambda) = \lambda \cdot x + (1 - \lambda) \cdot \bar{x} \in X,$$

and by the chain rule:

$$\lim_{\lambda \to 0+} \frac{f(z(\lambda)) - f(z(0))}{\lambda} = \nabla^T f(z(0)) \cdot \left.\frac{dz(\lambda)}{d\lambda}\right|_{\lambda=0} = \nabla^T f(\bar{x}) \cdot (x - \bar{x}) < 0.$$

However, if $\bar{x}$ is optimal, we have $f(z(\lambda)) \geq f(\bar{x}) = f(z(0))$ for all $\lambda \in [0, 1]$, a contradiction.

2.2.1.3 Prox-Function

Let $X \subset \mathbb{R}^n$ be a convex and closed set. We shall capture the geometry of X by means of an associated prox-function $d : X \to \mathbb{R}$. A *prox-function* should be continuously differentiable and *β-strongly convex* with respect to a norm $\| \cdot \|$, i.e. for all $x, y \in X$ it holds:

$$d(y) \geq d(x) + \nabla^T d(x) \cdot (y - x) + \frac{\beta}{2} \cdot \|y - x\|^2. \tag{$\mathcal{SC}^1$}$$

Note that the *convexity parameter β* measures the curvature of the prox-function d on X. In fact, if d happens to be twice continuously differentiable, its strong convexity can be characterized by the lower bound on the curvature of the graph, i.e. for all $x \in X$ and $\xi \in \mathbb{R}^n$ it holds:

$$\xi^T \cdot \nabla^2 f(x) \cdot \xi \geq \beta \cdot \|\xi\|^2. \tag{$\mathcal{SC}^2$}$$

For the proof of this assertion we refer to Nesterov (2018). Now, we turn our attention to the minimization of the prox-function:

$$\min_{x \in X} d(x). \tag{$\mathcal{D}$}$$

Due to the strong convexity of d, the optimization problem $(\mathcal{D})$ is uniquely solvable. Its unique solution is called the *prox-center* of X, i.e.

$$x(1) = \arg\min_{x \in X} d(x).$$

Another geometric notion for X in terms of d is its *diameter*

$$D = \sqrt{\max_{x,y \in X} d(x) - d(y)}.$$

2.2.1.4 Bregman Divergence

The prox-function d induces a distance-like function on X. This is the so-called *Bregman divergence* defined for all $x, y \in X$:

$$B(x, y) = d(x) - d(y) - \nabla^T d(y) \cdot (x - y).$$

Due to the strong convexity $(\mathcal{SC}^1)$ of d with exchanged $x, y \in X$, we have:

$$d(x) \geq d(y) + \nabla^T d(y) \cdot (x - y) + \frac{\beta}{2} \cdot \|x - y\|^2.$$

Thus, the Bregman divergence is nonnegative:

$$B(x, y) \geq \frac{\beta}{2} \cdot \|x - y\|^2.$$

The *three-points identity* of the Bregman divergence will be useful in what follows, see Exercise 2.3. The latter says that for all $x, y, z \in X$ it holds:

$$B(x, y) - B(x, z) - B(z, y) = \left(\nabla^T d(y) - \nabla^T d(z)\right) \cdot (z - x). \tag{$3\mathcal{PI}$}$$

With the Bregman divergence we further associate an auxiliary optimization problem:

$$\min_{x \in X} \ c^T \cdot x + B(x, y), \tag{$\mathcal{A}$}$$

where $y \in X$ and $c \in \mathbb{R}^n$ are given. Let us assume that $(\mathcal{A})$ is simple, i.e. it admits the unique closed-form solution.

2.2.1.5 Online Learning Technique

We make the following assumptions on the feasible set X and the loss functions f_t:

(A1) Let X be a convex and closed subset of $\mathbb{R}^n$ equipped with a prox-function $d : X \to \mathbb{R}$. We denote by β its convexity parameter with respect to a norm $\| \cdot \|$, and by $x(1)$ the corresponding prox-center of X. Let the diameter D of X in terms of d be finite. We associate with the prox-function d the Bregman divergence $B : X \times X \to \mathbb{R}$.

(A2) Let the loss functions $f_t : X \to \mathbb{R}$ be convex with uniformly bounded gradients ∇f_t with respect to the dual norm $\| \cdot \|_*$, i.e. there exists a positive constant G such that for all $x \in X$ and $t = 1, 2, \ldots$ it holds:

$$\| \nabla f_t(x) \|_* \leq G.$$

Now, we are ready to present the *online mirror descent*. Accordingly, the next decision minimizes the linearization of the loss function regularized by the Bregman divergence on the feasible set. For $t = 1, 2, \ldots$ we thus set:

$$x(t+1) = \underset{x \in X}{\arg\min} \; \underbrace{f_t\,(x(t)) + \nabla^T f_t\,(x(t)) \cdot x}_{\substack{\text{linearization of} \\ \text{loss function}}} \quad + \quad \underbrace{\frac{1}{\eta} \cdot B(x, x(t))}_{\substack{\text{regularization by} \\ \text{Bregman divergence}}}.$$

$$(\mathcal{OMD})$$

Here, $x(1)$ is the prox-center of X, and $\eta > 0$ is an appropriately chosen stepsize. Note that the $(\mathcal{OMD})$ update can be equivalently written by using the auxiliary optimization problem $(\mathcal{A})$:

$$x(t+1) = \underset{x \in X}{\arg\min} \; c^T \cdot x + B(x, y),$$

where

$$c = \eta \cdot \nabla f_t\,(x(t)), \quad y = x(t).$$

2.2.1.6 Convergence Analysis of $(\mathcal{OMD})$

We turn our attention to the worst-case convergence analysis of $(\mathcal{OMD})$ aiming to derive a bound on the corresponding regret. Our exposition follows mainly the ideas from Beck and Teboulle (2003).

Step 1 (Estimation of Losses) It holds for any $x \in X$ and $t = 1, 2, \ldots$:

$$f_t\,(x(t)) - f(x) \leq \frac{1}{\eta} \cdot (B(x, x(t)) - B(x, x(t+1))) + \frac{\eta}{2\beta} \cdot \| \nabla f_t(x(t)) \|_*^2.$$

Proof of Step 1 The convexity of f_t provides due to $(\mathcal{SC}^1)$:

$$f_t\,(x(t)) - f(x) \leq \nabla^T f_t\,(x(t)) \cdot (x(t) - x),$$

or, by enlarging the right-hand side:

$$f_t\left(x(t)\right) - f(x) \leq \underbrace{\nabla^T f_t\left(x(t)\right) \cdot \left(x(t+1) - x\right)}_{I} + \underbrace{\nabla^T f_t\left(x(t)\right) \cdot \left(x(t) - x(t+1)\right)}_{II}.$$

Let us first estimate expression I. From the optimality condition ($\mathcal{OPT}$) applied for the ($\mathcal{OMD}$) update we obtain:

$$\left(\nabla f_t\left(x(t)\right) + \frac{1}{\eta} \cdot \left(\nabla d(x(t+1)) - \nabla d(x(t))\right)\right)^T \cdot \left(x - x(t+1)\right) \geq 0,$$

or, equivalently:

$$\nabla^T f_t\left(x(t)\right) \cdot \left(x(t+1) - x\right) \leq \frac{1}{\eta} \cdot \left(\nabla d(x(t)) - \nabla d(x(t+1))\right)^T \cdot \left(x(t+1) - x\right).$$

The latter can be rewritten by using the three-point identity ($3\mathcal{PI}$):

$$\left(\nabla d(x(t)) - \nabla d(x(t+1))\right)^T \cdot \left(x(t+1) - x\right) = B(x, x(t)) - B(x, x(t+1)) - B(x(t+1), x(t)).$$

Overall, we obtain:

$$I \leq \frac{1}{\eta} \cdot \left(B(x, x(t)) - B(x, x(t+1)) - B(x(t+1), x(t))\right).$$

Now, we estimate expression II. The Hölder inequality ($\mathcal{H}$) provides:

$$\nabla^T f_t\left(x(t)\right) \cdot \left(x(t) - x(t+1)\right) \leq \|\nabla f_t\left(x(t)\right)\|_* \cdot \|x(t) - x(t+1)\|.$$

Moreover, due to

$$\left(\sqrt{\frac{\eta}{2\beta}} \cdot \|\nabla f_t\left(x(t)\right)\|_* - \sqrt{\frac{\beta}{2\eta}} \cdot \|x(t) - x(t+1)\|\right)^2 \geq 0,$$

we have by expanding and rearranging:

$$\|\nabla f_t\left(x(t)\right)\|_* \cdot \|x(t) - x(t+1)\| \leq \frac{\eta}{2\beta} \cdot \|\nabla f_t\left(x(t)\right)\|_*^2 + \frac{\beta}{2\eta} \cdot \|x(t) - x(t+1)\|^2.$$

Overall, we obtain:

$$II \leq \frac{\eta}{2\beta} \cdot \|\nabla f_t\left(x(t)\right)\|_*^2 + \frac{\beta}{2\eta} \cdot \|x(t+1) - x(t)\|^2.$$

We are done by comparing the estimations for I and II and recalling that

$$B(x(t+1), x(t)) \geq \frac{\beta}{2} \cdot \|x(t+1) - x(t)\|^2.$$

Step 2 (Bound on Regret) For any $T > 0$ it holds:

$$\mathcal{R}(T) \leq \frac{1}{\eta} \cdot \frac{D^2}{T} + \eta \cdot \frac{G^2}{2\beta}.$$

Proof of Step 2 For the average regret we have:

$$\mathcal{R}(T) = \frac{1}{T} \cdot \sum_{t=1}^{T} f_t(x(t)) - \min_{x \in X} \frac{1}{T} \cdot \sum_{t=1}^{T} f_t(x) = \frac{1}{T} \cdot \max_{x \in X} \sum_{t=1}^{T} (f_t(x(t)) - f_t(x)).$$

By using Step 1 and summing up over $t = 1, \ldots, T$, we obtain:

$$\sum_{t=1}^{T} (f_t(x(t)) - f_t(x)) \leq \underbrace{\frac{1}{\eta} \cdot \sum_{t=1}^{T} (B(x, x(t)) - B(x, x(t+1)))}_{III} + \underbrace{\frac{\eta}{2\beta} \cdot \sum_{t=1}^{T} \|\nabla f_t(x(t))\|_*^2}_{IV}.$$

For expression III we have:

$$\sum_{t=1}^{T} (B(x, x(t)) - B(x, x(t+1))) = B(x, x(1)) - \underbrace{B(x, x(T+1))}_{\geq 0} \leq B(x, x(1)).$$

Now, we recall that $x(1)$ is the prox-center of X, i.e. the unique solution of the optimization problem $(\mathcal{D})$. From the optimality condition $(\mathcal{OPT})$ for $(\mathcal{D})$ we obtain:

$$\nabla^T d\,(x(1)) \cdot (x - x(1)) \geq 0.$$

Hence, the definition of Bregman divergence provides:

$$B\,(x, x(1)) = d(x) - d(x(1)) - \underbrace{\nabla^T d(x(1)) \cdot (x - x(1))}_{\geq 0} \leq d(x) - d(x(1)) \leq D^2.$$

In order to estimate expression IV, we use the uniform bound on the gradients of loss functions:

$$\sum_{t=1}^{T} \|\nabla f_t(x(t))\|_*^2 \leq \sum_{t=1}^{T} G^2 = T \cdot G^2.$$

The comparison of estimates for III and IV concludes the proof of Step 2.

Step 3 (Adjustment of Stepsizes) For any $T > 0$ it holds:

$$\mathcal{R}(T) \leq D \cdot G \cdot \sqrt{\frac{2}{\beta \cdot T}}$$

by choosing the stepsize

$$\eta = \frac{D}{G} \cdot \sqrt{\frac{2\beta}{T}}.$$

Proof of Step 3 It is convenient to minimize the bound from Step 2 with respect to the stepsize η. The details are shown in Exercise 2.7. We conclude that the average regret $\mathcal{R}(T)$ is asymptotically vanishing for $T \rightarrow \infty$, namely, at the order $1/\sqrt{T}$. This is the optimal rate of convergence for online learning schemes, see e.g. Hazan (2016). We have shown that the decisions $x(t), t = 1, \ldots, T$, formed by means of $(\mathcal{OMD})$, achieve this rate.

2.2.2 Entropic Setup

We explicitly state the $(\mathcal{OMD})$ algorithm in the entropic setup. This enables us to efficiently solve the problem of portfolio selection in the online mode.

2.2.2.1 Manhattan and Maximum Norm

We consider *Manhattan* and *maximum norm* on $\mathbb{R}^n$:

$$\|x\|_1 = \sum_{i=1}^{n} |x_i|, \quad \|x\|_\infty = \max_{i=1,\ldots,n} |x_i|.$$

The Manhattan norm is the sum of the absolute values of the vector's entries, whereas the maximum norm is their maximum. It is easy to see that both functions $\|\cdot\|_1$ and $\|\cdot\|_\infty$ are positive definite, absolutely homogeneous, and satisfy the triangle inequality. Let us show that the Manhattan norm is dual of the maximum norm. First, we estimate:

$$g^T \cdot x \leq \sum_{i=1}^{n} |g_i| \cdot |x_i| \leq \max_{i=1,\ldots,n} |g_i| \cdot \sum_{i=1}^{n} |x_i| = \|g\|_\infty \cdot \|x\|_1.$$

Hence, the dual norm is upper bounded:

$$\max_{\|x\|_1 \leq 1} g^T \cdot x \leq \|g\|_\infty \cdot \|x\|_1 \leq \|g\|_\infty.$$

It remains to show that this upper bound is attained. For that, let i be an index with

$$\|g\|_\infty = \max_{i=1,\ldots,n} |g_i| = |g_i|.$$

By taking $\bar{x}_i = \text{sign}(g_i)$ and $\bar{x}_j = 0$ for all $j \neq i$, we obtain:

$$\|\bar{x}\|_1 = |\text{sign}(g_i)| = 1, \quad g^T \cdot \bar{x} = g_i \cdot \text{sign}(g_i) = |g_i| = \|g\|_\infty.$$

2.2.2.2 Loss Functions

We consider the logarithmic loss function

$$f_t(x) = -\ln r^T(t) \cdot x.$$

Let us compute its gradient and Hesse matrix by using the chain rule:

$$\nabla f_t(x) = -\frac{r(t)}{r^T(t) \cdot x}, \quad \nabla^2 f_t(x) = \frac{r(t) \cdot r^T(t)}{\left(r^T(t) \cdot x\right)^2}.$$

We see that its Hesse matrix is positive semidefinite, i.e. for all $x \in \Delta$ and $\xi \in \mathbb{R}^n$ it holds:

$$\xi^T \cdot \nabla f_t^2(x) \cdot \xi = \xi^T \cdot \frac{r(t) \cdot r^T(t)}{\left(r^T(t) \cdot x\right)^2} \cdot \xi = \left(\frac{r^T(t) \cdot \xi}{r^T(t) \cdot x}\right)^2 \geq 0.$$

Applying the second-order criterion $(\mathcal{C}^2)$, we deduce that that f is convex. Now, let us introduce an assumption on the vector of assets' returns for $t = 1, 2, \ldots$:

$$\rho_{\min} \leq \|r(t)\|_\infty \leq \rho_{\max},$$

where $\rho_{\min}, \rho_{\max} > 0$ are lower and upper bound, respectively. Then, the gradients of the loss function become uniformly bounded, i.e. for all $x \in \Delta$ and $t \in \mathbb{N}$ it holds:

$$\|\nabla f_t(x)\|_\infty = \left\|-\frac{r(t)}{r^T(t) \cdot x}\right\|_\infty = \frac{\|r(t)\|_\infty}{r^T(t) \cdot x} \leq \frac{\rho_{\max}}{\rho_{\min} \cdot \sum_{i=1}^n x_i} = \frac{\rho_{\max}}{\rho_{\min}} = G.$$

2.2.2.3 Negative Entropy

We consider the *negative entropy* on the simplex:

$$d(x) = \sum_{i=1}^n x_i \cdot \ln x_i, \quad x \in \Delta.$$

First, we see that d is continuously differentiable with partial derivatives:

$$\frac{\partial d}{\partial x_i}(x) = \ln x_i + \frac{x_i}{x_i} = \ln x_i + 1, \quad i = 1, \ldots, n.$$

Its Hesse matrix is diagonal:

$$\nabla^2 d(x) = \mathrm{diag}\left(\frac{1}{x_1}, \ldots, \frac{1}{x_n}\right).$$

We show that d is 1-strongly convex with respect to the Manhattan norm. For that, we apply the second-order criterion $(\mathcal{SC}^2)$ for strong convexity with $\beta = 1$ and $\|\cdot\| = \|\cdot\|_1$:

$$\xi^T \cdot \nabla^2 d(x) \cdot \xi = \sum_{i=1}^{n} \frac{\xi_i^2}{x_i} \geq \left(\sum_{i=1}^{n} |\xi_i|\right)^2 = \|\xi\|_1^2.$$

The last inequality holds due to

$$\sum_{i=1}^{n} |\xi_{(i)}| = \sum_{i=1}^{n} \frac{|\xi_{(i)}|}{\sqrt{x_{(i)}}} \cdot \sqrt{x_{(i)}} \leq \sqrt{\sum_{i=1}^{n} \frac{\xi_{(i)}^2}{x_{(i)}}} \cdot \sqrt{\sum_{i=1}^{n} x_{(i)}} = \sqrt{\sum_{i=1}^{n} \frac{\xi_{(i)}^2}{x_{(i)}}}.$$

The strong convexity of the *entropic prox-function* can be alternatively shown directly by using the Pinsker inequality from information theory. The computation of the prox-center, as well as the estimation of the diameter in the entropic setup is postponed to Exercise 2.4:

$$x(1) = \left(\frac{1}{n}, \ldots, \frac{1}{n}\right)^T, \quad D \leq \sqrt{\ln n}.$$

2.2.2.4 Kullback-Leibler Divergence

We derive the Bregman divergence corresponding to the entropic prox-function on the simplex:

$$B(x, y) = d(x) - d(y) - \nabla^T d(y) \cdot (x - y)$$

$$= \sum_{i=1}^{n} x_i \cdot \ln x_i - \sum_{i=1}^{n} y_i \cdot \ln y_i - \sum_{i=1}^{n} (\ln y_i + 1) \cdot (x_i - y_i)$$

$$= \sum_{i=1}^{n} x_i \cdot \ln x_i - \sum_{i=1}^{n} x_i \cdot \ln y_i = \sum_{i=1}^{n} x_i \cdot \ln \frac{x_i}{y_i}.$$

This is the *Kullback-Leibler divergence* from information theory. It remains to solve the auxiliary optimization problem ($\mathcal{A}$) on the simplex:

$$\min_{x \geq 0} \; c^T \cdot x + B(x, y) = \sum_{i=1}^{n} c_i \cdot x_i + \sum_{i=1}^{n} x_i \cdot \ln \frac{x_i}{y_i} \quad \text{s.t.} \quad \sum_{i=1}^{n} x_i - 1 = 0,$$

where $y \in \Delta$ and $c \in \mathbb{R}^n$ are given. By introducing a multiplier $\mu \in \mathbb{R}$ for the equality constraint, the Lagrange multiplier rule reads componentwise as follows, see e.g. Jongen et al. (2004):

$$c_i + \ln \frac{x_i}{y_i} + 1 = \mu,$$

or rearranged:

$$x_i = e^{\mu-1} \cdot y_i \cdot e^{-c_i}.$$

Summing up over $i = 1, \ldots, n$, we have:

$$\sum_{i=1}^{n} x_i = e^{\mu-1} \cdot \sum_{i=1}^{n} y_i \cdot e^{-c_i}.$$

Recalling $e^T \cdot x = 1$, it follows:

$$e^{\mu-1} = \frac{1}{\displaystyle\sum_{i=1}^{n} y_i \cdot e^{-c_i}}.$$

Substituting back, we finally obtain the unique solution of ($\mathcal{A}$):

$$x_i = \frac{y_i \cdot e^{-c_i}}{\displaystyle\sum_{i=1}^{n} y_i \cdot e^{-c_i}}, \quad i = 1, \ldots, n.$$

2.2.2.5 Online Portfolio Selection

We apply ($\mathcal{OMD}$) in the context of portfolio selection. For $t = 1, 2, \ldots$ we set:

$$x_i(t+1) = \frac{x_i(t) \cdot e^{\frac{\eta \cdot r_i(t)}{r^T(t) \cdot x(t)}}}{\displaystyle\sum_{i=1}^{n} x_i(t) \cdot e^{\frac{\eta \cdot r_i(t)}{r^T(t) \cdot x(t)}}}, \quad i = 1, \ldots, n.$$

Here, we start with the *equally weighted portfolio*:

$$x(1) = \left(\frac{1}{n}, \ldots, \frac{1}{n}\right)^T.$$

As soon as the returns $r(t)$ become known, the next portfolio $x(t + 1)$ is realized. Note that this new portfolio $x(t + 1)$ depends not only on the current iterate $x(t)$, but also on

$$\frac{r_i(t)}{r^T(t) \cdot x(t)}, \quad i = 1, \ldots, n.$$

The latter quantities represent the shares of the i-th asset's return $r_i(t)$, $i = 1, \ldots, n$ within the portfolio's return $r^T(t) \cdot x(t)$. Further, let us specify the stepsize:

$$\eta = \frac{\rho_{\min}}{\rho_{\max}} \cdot \sqrt{\frac{2 \ln n}{T}}.$$

The convergence analysis of $(\mathcal{OMD})$ provides:

$$\mathcal{R}(T) \leq \frac{\rho_{\max}}{\rho_{\min}} \cdot \sqrt{\frac{2 \ln n}{T}}.$$

The average regret $\mathcal{R}(T)$ is shown to asymptotically vanish as $T \to \infty$.

2.3   Case Study: Expert Advice

We present a model of prediction based on *expert advice*. Historically, this is the first framework proposed for online learning. To be specific, let a company be supplying a homogeneous good S and deciding repeatedly upon its retail price p. The corresponding demand D is hardly to be estimated and is varying over time. This is actually due to competitors' interventions on the market and its possible volatility. In order to cope with demand uncertainty, the company asks for expert advice on the pricing policy. One can easily assume that the marketing experts have access to external sources of information to make their decisions. There are n experts, each of them proposing a to be optimal price p_i from an interval $[0, \bar{p}]$. The company then predicts the average price

$$p^T \cdot x = \sum_{i=1}^{n} p_i \cdot x_i,$$

where $x \in \Delta$ models the distribution of expert credibility, i.e.

$$x_i \geq 0, \, i = 1, \ldots, n, \quad \sum_{i=1}^{n} x_i = 1.$$

The expert credibility x has to be adjusted by maximizing the company's sales revenue

$$\left(p^T \cdot x \right) \cdot \min \, \{D, S\}.$$

Task 1 State the problem of expert advice within the framework of online learning. How does the specification of the average regret look like?

Hint 1 The average regret is

$$\mathcal{R}(T) = \frac{1}{T} \cdot \sum_{t=1}^{T} f_t(x(t)) - \min_{x \in X} \frac{1}{T} \cdot \sum_{t=1}^{T} f_t(x),$$

where the feasible set and the loss functions are as follows:

$$X = \Delta, \quad f_t(x) = - \left(p^T(t) \cdot x \right) \cdot \min \, \{D(t), S\}.$$

Task 2 Interpret the average regret in economic terms. Show that the goal of credibility adjustment is to perform on the sales revenue as well as the best expert, i.e.

$$\mathcal{R}(T) = \max_{i=1,\ldots,n} \frac{1}{T} \cdot \sum_{t=1}^{T} p_i(t) \cdot \min \, \{D(t), S\} - \frac{1}{T} \cdot \sum_{t=1}^{T} \left(p^T(t) \cdot x(t) \right) \cdot \min \, \{D(t), S\}.$$

Hint 1 Derive and subsequently use the following identity for a given vector $c \in \mathbb{R}^n$:

$$\max_{x \in \Delta} c^T \cdot x = \max_{i=1,\ldots,n} c_i.$$

Task 3 Apply $(\mathcal{OMD})$ for the problem of expert advice in the entropic setup.

Hint 3 $(\mathcal{OMD})$ in the entropic setup reads for $t = 1, 2, \ldots$ as follows:

$$x_i(t+1) = \frac{x_i(t) \cdot e^{\eta \cdot p_i(t) \cdot \min \, \{D(t), S\}}}{\sum_{i=1}^{n} x_i(t) \cdot e^{\eta \cdot p_i(t) \cdot \min \, \{D(t), S\}}}, \quad i = 1, \ldots, n.$$

Here, we start with the equally weighted credibility $x(1) = \left(\frac{1}{n}, \ldots, \frac{1}{n}\right)^T$.

Task 4 By an appropriate specification of the stepsize η in the $(\mathcal{OMD})$ update, derive an optimal rate of convergence for the corresponding average regret $\mathcal{R}(T)$.

Hint 4 It holds for any $T > 0$:

$$\mathcal{R}(T) \leq S \cdot \bar{p} \cdot \sqrt{\frac{2 \ln n}{T}}.$$

2.4   Exercises

Exercise 2.1 (Dual Norm) Let $\|\cdot\|$ be an arbitrary norm on $\mathbb{R}^n$. Show that the corresponding dual norm is positive definite, absolutely homogeneous, and satisfies the triangle inequality:

$$\|g\|_* = \max_{\|x\| \leq 1} g^T \cdot x.$$

Exercise 2.2 (Cauchy-Schwarz Inequality) Show that the *Euclidean norm* $\|\cdot\|_2$ is self-dual:

$$\|x\|_2 = \sqrt{\sum_{i=1}^{n} x_i^2}.$$

Derive the *Cauchy-Schwarz inequality*:

$$\left| g^T \cdot x \right| \leq \|g\|_2 \cdot \|x\|_2. \tag{$\mathcal{CS}$}$$

Exercise 2.3 (Three-Points Identity) Let d be a prox-function on $X \subset \mathbb{R}^n$. Prove the third-points identity $(3\mathcal{PI})$ for the induced Bregman divergence, i.e. for all $x, y, z \in X$ it holds:

$$B(x, y) - B(x, z) - B(z, y) = \left(\nabla^T d(y) - \nabla^T d(z)\right) \cdot (z - x).$$

Exercise 2.4 (Negative Entropy) Consider the entropic prox-function on the simplex:

$$d(x) = \sum_{i=1}^{n} x_i \cdot \ln x_i, \quad x \in \Delta.$$

Compute the corresponding prox-center of the simplex and estimate its diameter:

$$x(1) = \left(\frac{1}{n}, \ldots, \frac{1}{n} \right)^T, \quad D \leq \sqrt{\ln n}.$$

Exercise 2.5 (Euclidean Setup) Elaborate the Euclidean setup for online learning:

(i) Show that $d(x) = \frac{1}{2} \cdot \|x\|_2^2$ is a prox-function with respect to the Euclidean norm $\| \cdot \|_2$, and its convexity parameter is $\beta = 1$. We call it the *Euclidean prox-function*.
(ii) Show that $B(x, y) = \frac{1}{2} \cdot \|x - y\|_2^2$ is the Bregman divergence induced by the Euclidean prox-function d. We call it the *Euclidean divergence*.

Exercise 2.6 (Projection) Let $X \subset \mathbb{R}^n$ be a closed and convex set. Consider the auxiliary optimization problem $(\mathcal{A})$ equipped with the Euclidean divergence:

$$\min_{x \in X} \; c^T \cdot x + \frac{1}{2} \cdot \|x - y\|_2^2,$$

where $y \in X$ and $c \in \mathbb{R}^n$ are given. Show that the latter is equivalent to the projection problem:

$$\min_{x \in X} \; \|x - (y - c)\|_2.$$

Its unique solution is called the *Euclidean projection* of $y - c$ on X, i.e.

$$\text{proj}_X (y - c) = \arg\min_{x \in X} \; \|x - (y - c)\|_2.$$

Exercise 2.7 (Online Gradient Descent, See Zinkevich (2003)) Let $X \subset \mathbb{R}^n$ be a closed and convex set. Let further its diameter D with respect to the Euclidean prox-function be finite. Additionally, assume that the loss functions $f_t : X \rightarrow \mathbb{R}$ are continuously differentiable and convex with a uniform bound G on their gradients ∇f_t with respect to the Euclidean norm. Show that $(\mathcal{OMD})$ in the Euclidean setup boils down to the so-called *online gradient descent*:

$$x(t + 1) = \text{proj}_X (x(t) - \eta \cdot \nabla f_t(x(t))), \quad x(1) = \text{proj}_X(0). \qquad (\mathcal{OGD})$$

By an appropriate specification of the stepsize η, derive an optimal rate of convergence for the corresponding average regret, i.e. for any $T > 0$ it holds:

$$\mathcal{R}(T) \leq D \cdot G \cdot \sqrt{\frac{2}{T}}.$$

Recommendation Systems

The purpose of a *recommendation system* is to predict how strong a user's interest is in not yet consumed products. The user is then offered the most attractive product according to the prediction. Typical recommendation services include videos and movies as for Netflix, YouTube, and Spotify, consumption goods as for Amazon, pieces of social content as for Facebook and Twitter. Recommendation systems are supposed to contribute to the management of the information overload by suggesting a relevant subset from an unmanageable amount of products to the user. Aiming to generate appropriate predictions, mathematical methods of *information retrieval* are used. In this chapter, we discuss *collaborative filtering* techniques and apply them for the prediction of *movie ratings*, and for the analysis of *latent semantics* in the documents. The neighborhood- and model-based approaches of collaborative filtering are elaborated in detail. Within the neighborhood-based approach, *similarity measures* are introduced and the *k-nearest neighbors* algorithm is described. The model-based approach uses a linear-algebraic technique of *singular value decomposition*. Singular value decomposition allows to reveal hidden patterns of users' choice behavior. After imposing a low-rank model on the latter, the prediction becomes optimization-driven. For solving the corresponding *low-rank approximation* problem, we apply the well-known optimization algorithm of gradient descent. An efficient application of gradient descent is enabled by matrix factorization of the low-rank approximation.

3.1 Motivation: Netflix Prize

In 2006, Netflix has called on the best recommendation system to predict user ratings for movies, based on previously specified ratings. This competition, also known as *Netflix prize*, attracted a lot of attention over the coming years and has contributed significantly to the data science research community. The main obstacle was that no other

© Springer-Verlag GmbH Germany, part of Springer Nature 2021
V. Shikhman, D. Müller, *Mathematical Foundations of Big Data Analytics*,
https://doi.org/10.1007/978-3-662-62521-7_3

information about the users or movies, except of their ratings, has been provided. As Netflix movie ratings were released, the task was to predict missing ratings of particular user-movie combinations. For this purpose, Netflix provided a vast data set containing 100,480,507 ratings that 480,189 users gave to 17,770 movies. The qualifying data set contained 2,817,131 user-movie combinations of missing ratings, known only to the jury. A participating team's algorithm must have predicted missing ratings on the entire qualifying set. In 2009, Netflix announced team "BellKor's Pragmatic Chaos" as the prize winner. They achieved an improvement of slightly more than 10 % over the Netflix's own benchmark algorithm, called Cinematch, on the qualifying set.

Let us mathematically model the Netflix prize problem. A *recommendation system* relies on ratings of m movies previously specified by n users. Those are stored in a *rating* $(n \times m)$-*matrix* $R = (r_{ij})$. Its entry r_{ij} in the i-th row and j-th column expresses the rating of the movie j by the user i. Exemplarily, let the following Netflix matrix with ratings from 1 to 5 be given:

$$
R = \begin{pmatrix}
 & \boxed{\text{M1}} & \boxed{\text{M2}} & \boxed{\text{M3}} & \boxed{\text{M4}} \\
\boxed{\text{U1}} & 5 & 3 & - & 1 \\
\boxed{\text{U2}} & 4 & - & - & 1 \\
\boxed{\text{U3}} & 1 & 1 & - & 5 \\
\boxed{\text{U4}} & 1 & - & - & 4 \\
\boxed{\text{U5}} & - & 1 & 5 & 4
\end{pmatrix}.
$$

The i-th row of R consists of ratings specified by the i-th user for an every movie. The j-th column of R consists of j-th movie's ratings specified by the users. Note that an entry r_{ij} is missing if the i-th user has not yet specified the j-th movie. The main goal of a recommendation system is to reasonably complete the missing ratings by taking into account the specified ones. This task is traditionally addressed by analysing user-item interactions. The corresponding prediction procedure is referred to as *collaborative filtering*. Collaborative filtering uses the idea that, although the ratings were specified by multiple users individually, their choice behavior underlies certain collectively validated patterns. The main challenge in designing collaborative filtering methods is that the underlying rating matrix R is *sparse*. In fact, most users would have viewed only a small fraction of the large universe of available movies. As a result, most of the ratings are missing. Nevertheless, if a user shows up and asks for a movie recommendation, we should be able to provide an appealing option.

3.2 Results

3.2.1 Neighborhood-Based Approach

A straightforward and, actually, the earliest attempt to collaborative filtering is *neighborhood-based* approach, see e.g. Aggarwal (2016). The latter postulates that similar users display similar patterns of rating behavior and similar items receive similar ratings. The basic idea is to determine users, who are similar to the target user, and recommend for the corresponding missing ratings just the weighted averages of the ratings of this peer group. Let us cast this idea in mathematical terms.

3.2.1.1 Similarity Measures

First, let us introduce similarity measures $\mathrm{Sim}(i, \ell)$ between users i and ℓ. For that, we consider those movies for which ratings have been specified by the i-th user:

$$M_i = \left\{ j \in \{1, \ldots m\} \mid r_{ij} \text{ is specified} \right\}.$$

For two users i and ℓ the set $M_i \cap M_\ell$ of mutually specified movies is used to compute the similarity between them. The *cosine similarity* between the users i and ℓ is defined as

$$\mathrm{Cosine}(i, \ell) = \frac{\sum\limits_{j \in M_i \cap M_\ell} r_{ij} \cdot r_{\ell j}}{\sqrt{\sum\limits_{j \in M_i \cap M_\ell} r_{ij}^2} \cdot \sqrt{\sum\limits_{j \in M_i \cap M_\ell} r_{\ell j}^2}}.$$

The cosine similarity measures the angle between the corresponding rating vectors:

$$r_i = \left(r_{ij}, j \in M_i \cap M_\ell \right)^T, \quad r_\ell = \left(r_{\ell j}, j \in M_i \cap M_\ell \right)^T.$$

By using the scalar product and the Euclidean norm, it holds namely:

$$\mathrm{Cosine}(i, \ell) = \frac{r_i^T \cdot r_\ell}{\|r_i\|_2 \cdot \|r_\ell\|_2}.$$

Another popular similarity measure is the Pearson correlation coefficient. For that, we define by averaging the mean ratings corresponding to the users i and ℓ:

$$\mu_i = \frac{1}{|M_i \cap M_\ell|} \cdot \sum_{j \in M_i \cap M_\ell} r_{ij}, \quad \mu_\ell = \frac{1}{|M_i \cap M_\ell|} \cdot \sum_{j \in M_i \cap M_\ell} r_{\ell j},$$

where we denote by $|M_i \cap M_\ell|$ the number of mutually specified movies. The *Pearson correlation coefficient* between the users i and ℓ is defined as

$$\text{Pearson}(i, \ell) = \frac{\sum\limits_{j \in M_i \cap M_\ell} \left(r_{ij} - \mu_i\right) \cdot \left(r_{\ell j} - \mu_\ell\right)}{\sqrt{\sum\limits_{j \in M_i \cap M_\ell} \left(r_{ij} - \mu_i\right)^2} \cdot \sqrt{\sum\limits_{j \in M_i \cap M_\ell} \left(r_{\ell j} - \mu_\ell\right)^2}}.$$

The latter can be equivalently rewritten in statistical terms:

$$\text{Pearson}(i, \ell) = \frac{\sigma_{i\ell}}{\sigma_i \cdot \sigma_\ell},$$

where $\sigma_{i\ell}$ stands for the covariance of rating vectors r_i and r_ℓ, whereas σ_i, σ_ℓ for their standard deviations, respectively. In contrast to cosine similarity, Pearson correlation coefficient is not biased with respect to mean ratings. This adjustment accounts for the fact that users exhibit different levels of generosity in their choice behavior.

3.2.1.2 *k*-Nearest Neighbors

Let a similarity measure $\text{Sim}(i, \ell)$ between the users i and ℓ be chosen—it can be either cosine similarity or Pearson correlation coefficient. Based on that, we introduce the *k*-nearest neighbors algorithm for recommendation. The *k*-nearest neighbors algorithm works much in the way some of us asks for movie recommendations from our friends. First, we start with users whose taste we feel we share, and then we ask a bunch of them— these are our "neighbors"—to recommend something to us. If many of them recommend the same thing, we deduce that we shall like it as well. In what follows, we try to mathematically formalize this procedure. For that, we assume that the i-th user has not yet specified the rating of the j-th movie, so that the rating r_{ij} is just missing. We consider those users who specified their ratings of the j-th movie:

$$U_j = \left\{\ell \in \{1, \ldots n\} \mid r_{\ell j} \text{ is specified}\right\}.$$

We assume that the set U_j is not empty, and denote by $n_j = |U_j| > 0$ its cardinality. Let us sort the users from U_j in decreasing order of their similarity to the i-th user, i.e.

$$U_j = \left\{\ell_1, \ldots, \ell_{n_j}\right\} \quad \text{with} \quad \text{Sim}\left(i, \ell_1\right) \geq \ldots \geq \text{Sim}\left(i, \ell_{n_j}\right).$$

Let us now select at most k nearest neighbors of the i-th user:

$$N_j(i) = \left\{\ell_1, \ldots, \ell_{\min\{k, n_j\}}\right\}.$$

Finally, the missing rating of the j-th movie is set as the weighted sum of the corresponding ratings specified by k-nearest neighbors of the i-th user:

$$r_{ij} = \sum_{\ell \in N_j(i)} \frac{\text{Sim}(i, \ell)}{\sum_{\ell \in N_j(i)} |\text{Sim}(i, \ell)|} \cdot r_{\ell j}. \qquad (k\mathcal{N}\mathcal{N})$$

Let us apply the $(2\mathcal{N}\mathcal{N})$ algorithm to the Netflix matrix R from above. We impose the cosine similarity measure for users. The neighborhood-based completion is as follows, cf. Exercise 3.1:

$$R_{neighbor} = \begin{pmatrix} & \boxed{\text{M1}} & \boxed{\text{M2}} & \boxed{\text{M3}} & \boxed{\text{M4}} \\ \boxed{\text{U1}} & 5 & 3 & \mathbf{5.00} & 1 \\ \boxed{\text{U2}} & 4 & \mathbf{1.99} & \mathbf{5.00} & 1 \\ \boxed{\text{U3}} & 1 & 1 & \mathbf{5.00} & 5 \\ \boxed{\text{U4}} & 1 & \mathbf{1.00} & \mathbf{5.00} & 4 \\ \boxed{\text{U5}} & \mathbf{2.50} & 1 & 5 & 4 \end{pmatrix}.$$

The advantages of neighborhood-based techniques are that they are simple to implement and the resulting recommendations are often easy to explain. However, as we see from our example, $(k\mathcal{N}\mathcal{N})$ methodology has also some drawbacks. First, sparsity creates challenges for robust similarity computation when the number of mutually rated movies between two users is small. In Netflix application, this will be unfortunately rather often the case. As illustration, movie M3 has the same recommended rating for users U1–U4 just because of the user's U5 specified rating. Second, $(k\mathcal{N}\mathcal{N})$ does not reveal common patterns behind the users' choice behavior. In fact, the prediction phase is user-specific: the neighbors of a user need to be determined before prediction starts. There is no model specifically created up front for prediction rather than we have a preprocessing phase, which is required to ensure efficient implementation. To overcome this fundamental drawback requires an alternative approach to collaborative filtering.

3.2.2 Model-Based Approach

Model-based collaborative filtering uses the specified rating data to estimate a prediction model, see e.g. Aggarwal (2016). The model is created up front, i.e. the modeling phase is clearly separated from the prediction phase. In Netflix application, the modeling is based on the idea that users focus just on several important features while specifying movie ratings. Obviously, nearly all users rate movies according to a more or less uniform evaluation scheme. Be it film genres, famous actors involved, special shooting locations, exciting movie scripts, prominent directors, etc.—there are only a few characteristics that are relevant for all the ratings given. This point of view allows to incorporate structural properties of users' choice behavior into the model in an ad hoc manner. In the subsequently launched prediction phase, the parameters of the underlying model are learned by means of optimization techniques.

3.2.2.1 Singular Value Decomposition

In order to build up a prediction model, let us assume for a moment that all entries of the rating $(n \times m)$-matrix R are specified. How is it possible to reveal hidden patterns behind these ratings? In what follows, we want to identify features of m movies, upon which n users specify their ratings. This is done by means of a linear-algebraic technique called singular value decomposition, see e.g. Lancaster (1969). For the rating matrix R we present the *singular value decomposition* in reduced form:

$$R = U \cdot \Sigma \cdot V. \qquad (\mathcal{SVD})$$

Here, $U = (u_{ik})$ is an $(n \times r)$-matrix with orthogonal columns, and $V = (v_{kj})$ is an $(r \times m)$-matrix with orthogonal rows. Denoting by I the identity matrix of the right size, the latter reads:

$$U^T \cdot U = I, \quad V \cdot V^T = I.$$

Σ is the diagonal $(r \times r)$-matrix, i.e.

$$\Sigma = \text{diag} (\sigma_1, \ldots, \sigma_r) .$$

Its diagonal entries σ_i, $i = 1, \ldots, r$, are called *singular values* and turn out to be positive. Without loss of generality, we sort them in nondecreasing order:

$$\sigma_1 \geq \ldots \geq \sigma_r > 0.$$

The number r of positive singular values corresponds to the rank of matrix R, i.e.

$$r = \text{rank}(R).$$

Let us interpret the singular value decomposition in terms of *latent features*. U can be viewed as a matrix of user-feature ratings. Its entry u_{ik} is the rating of the k-th feature specified by the i-th user. V can be viewed as a matrix of feature-movie characteristics. Its entry v_{kj} characterizes the manifestation of the k-th feature in the j-th movie. Singular value σ_k expresses the importance of the k-th feature for all users collectively. By multiplying out the matrix product in $(\mathcal{SVD})$, we get the rating of the j-th movie specified by the i-th user as the weighted sum:

$$r_{ij} = \sum_{k=1}^{r} \sigma_k \cdot u_{ik} \cdot v_{kj},$$

where the number of features is given by r. Overall, latent features reveal the hidden patterns behind the users' choice behavior.

3.2.2.2 Left- and Right-Singular Vectors

Let us start with the derivation of $(\mathcal{SVD})$. We first show the existence of *left- and right-singular vectors* $u_i \in \mathbb{R}^n$, $v_i \in \mathbb{R}^m$, $i = 1, \ldots, \min\{n, m\}$, of the matrix R, respectively. They should fulfill:

$$R \cdot v_i = \sigma_i \cdot u_i, \quad R^T \cdot u_i = \sigma_i \cdot v_i,$$

and be pairwise orthogonal:

$$u_i^T \cdot u_j = v_i^T \cdot v_j = \begin{cases} 1, \text{ if } i = j, \\ 0, \text{ if } i \neq j. \end{cases}$$

Let k pairs of left- and right-singular vectors u_i, v_i, $i = 1, \ldots, k$ be given. How is it possible to construct a $(k+1)$-st pair? For that, we apply the following variational principle:

$$\max_{u, v \in \mathbb{R}^n} \sigma(u, v) = u^T \cdot R \cdot v \quad \text{s.t.} \quad \begin{cases} \text{(i) } u^T \cdot u = v^T \cdot v = 1, \\ \text{(ii) } u_i^T \cdot u = v_i^T \cdot v = 0 \quad \text{for all } i = 1, \ldots, k. \end{cases}$$

By solving this optimization problem, we hope to find new left- and right-singular vectors u, v which are, on the one hand, orthogonal to the k previous ones, and maximize the singular value $\sigma(u, v)$, on the other hand. Note that the largest value of $\sigma(u, v)$ must be nonnegative. If it were negative, changing the sign of either u or v would make it positive and therefore larger. Further, we introduce multipliers $\lambda, \mu \in \mathbb{R}$ for the equality constraints (i), and multipliers $\lambda_i, \mu_i \in \mathbb{R}$ for the equality constraints (ii), $i = 1, \ldots, k$. The corresponding Lagrange multiplier rule reads as follows, see e.g. Jongen et al. (2004):

$$\nabla \sigma(u, v) = \lambda \cdot \nabla \left(u^T \cdot u \right) + \mu \cdot \nabla \left(v^T \cdot v \right) + \sum_{i=1}^{k} \lambda_i \cdot \nabla \left(u_i^T \cdot u \right) + \sum_{i=1}^{k} \mu_i \cdot \left(\nabla v_i^T \cdot v \right).$$

By computing the gradients, we obtain:

$$\text{(a) } R \cdot v = 2 \cdot \lambda \cdot u + \sum_{i=1}^{k} \lambda_i \cdot u_i, \quad \text{(b) } R^T \cdot u = 2 \cdot \mu \cdot v + \sum_{i=1}^{k} \mu_i \cdot v_i.$$

We multiply equation (a) by u^T, and equation (b) by v^T:

$$\underbrace{u^T \cdot R \cdot v}_{=\sigma(u,v)} = 2 \cdot \lambda \cdot \underbrace{u^T \cdot u}_{=1} + \sum_{i=1}^{k} \lambda_i \cdot \underbrace{u^T \cdot u_i}_{=0}, \quad \underbrace{v^T \cdot R^T \cdot u}_{=\sigma(u,v)} = 2 \cdot \mu \cdot \underbrace{v^T \cdot v}_{=1} + \sum_{i=1}^{k} \mu_i \cdot \underbrace{v^T \cdot v_i}_{=0}.$$

Hence, we get for the both Lagrange multipliers:

$$\sigma(u, v) = 2 \cdot \lambda = 2 \cdot \mu.$$

Next, we multiply equation (a) by u_i^T, and equation (b) by v_i^T for $i = 1, \ldots, k$:

$$u_i^T \cdot R \cdot v = 2 \cdot \lambda \cdot u_i^T \cdot u + u_i^T \cdot \left(\sum_{i=1}^{k} \lambda_i \cdot u_i \right), \quad v_i^T \cdot R^T \cdot u = 2 \cdot \mu \cdot v_i^T \cdot v + v_i^T \cdot \left(\sum_{i=1}^{k} \mu_i \cdot v_i \right).$$

After simplifying by means of (i) and (ii), we obtain:

$$u_i^T \cdot R \cdot v = \lambda_i, \quad v_i^T \cdot R^T \cdot u = \mu_i.$$

We use the fact that u_i, v_i are left- and right-singular vectors of R, respectively:

$$\lambda_i = u_i^T \cdot R \cdot v = \underbrace{\sigma_i \cdot v_i^T}_{=\sigma_i \cdot v_i^T} \cdot \underbrace{v}_{=0} = 0, \quad \mu_i = v_i^T \cdot R^T \cdot u = \underbrace{\sigma_i \cdot u_i^T}_{=\sigma_i \cdot u_i^T} \cdot \underbrace{u}_{=0} = 0.$$

Hence, we get for the other Lagrange multipliers:

$$\lambda_i = \mu_i = 0.$$

Overall, let us substitute the derived Lagrange multipliers into equations (a) and (b):

$$R \cdot v = \sigma(u, v) \cdot u, \quad R^T \cdot u = \sigma(u, v) \cdot v.$$

Finally, the next $(k + 1)$-st pair of left- and right-singular vectors is found by setting:

$$u_{k+1} = u, \quad v_{k+1} = v, \quad \sigma_{k+1} = \sigma(u, v).$$

Note that at most $\min\{n, m\}$ pairs of left- and right singular vectors can be constructed in such a way. Indeed, due to dimensionality reasons the feasible set of the above optimization problem remains empty if we try to obtain the $(\min\{n, m\} + 1)$-st pair.

3.2.2.3 Reduction

Let us write the equations defining left- and right-singular vectors in matrix form:

$$R \cdot \mathbf{V}^T = \mathbf{U} \cdot \mathbf{\Sigma}, \quad R^T \cdot \mathbf{U} = \mathbf{V}^T \cdot \mathbf{\Sigma},$$

where

$$\mathbf{U}^T \cdot \mathbf{U} = \mathbf{V} \cdot \mathbf{V}^T = I.$$

We distinguish the following cases:

- $m \leq n$ with the notation

$$\mathbf{U} = (u_1, \ldots, u_m), \quad \mathbf{V} = (v_1, \ldots, v_m)^T, \quad \mathbf{\Sigma} = \mathrm{diag}\,(\sigma_1, \ldots, \sigma_m).$$

Since the rows of $\mathbf{V}$ are orthogonal, and, in particular, linearly independent, the $(m \times m)$-matrix $\mathbf{V}$ is regular. For its inverse we obtain:

$$\mathbf{V}^{-1} = \mathbf{V}^{-1} \cdot \underbrace{\mathbf{V} \cdot \mathbf{V}^T}_{=I} = \underbrace{\mathbf{V}^{-1} \cdot \mathbf{V}}_{=I} \cdot \mathbf{V}^T = \mathbf{V}^T.$$

Hence, the inverse of $\mathbf{V}$ coincides with its transposition. Overall, we have shown that the matrix $\mathbf{V}$ is *orthogonal*, i.e.

$$\mathbf{V} \cdot \mathbf{V}^T = \mathbf{V}^T \cdot \mathbf{V} = I.$$

Now, we are ready to derive a singular value decomposition:

$$R = R \cdot \underbrace{\mathbf{V}^T \cdot \mathbf{V}}_{=I} = \underbrace{R \cdot \mathbf{V}^T}_{=\mathbf{U} \cdot \mathbf{\Sigma}} \cdot \mathbf{V} = \mathbf{U} \cdot \mathbf{\Sigma} \cdot \mathbf{V}.$$

- $m > n$ with the notation

$$\mathbf{U} = (u_1, \ldots, u_n), \quad \mathbf{V} = (v_1, \ldots, v_n)^T, \quad \mathbf{\Sigma} = \mathrm{diag}\,(\sigma_1, \ldots, \sigma_n).$$

Since the columns of $\mathbf{U}$ are orthogonal, and, in particular, linearly independent, the $(n \times n)$-matrix $\mathbf{U}$ is regular. For its inverse we obtain:

$$\mathbf{U}^{-1} = \underbrace{\mathbf{U}^T \cdot \mathbf{U}}_{=I} \cdot \mathbf{U}^{-1} = \mathbf{U}^T \cdot \underbrace{\mathbf{U} \cdot \mathbf{U}^{-1}}_{=I} = \mathbf{U}^T.$$

Hence, the inverse of $\mathbf{U}$ coincides with its transposition. Overall, we have shown that the matrix $\mathbf{U}$ is *orthogonal*, i.e.

$$\mathbf{U}^T \cdot \mathbf{U} = \mathbf{U} \cdot \mathbf{U}^T = I.$$

Now, we are ready to derive a singular value decomposition:

$$R = \underbrace{\mathbf{U} \cdot \mathbf{U}^T}_{=I} \cdot R = \mathbf{U} \cdot \left(\underbrace{R^T \cdot \mathbf{U}}_{=\mathbf{V}^T \cdot \mathbf{\Sigma}} \right)^T = \mathbf{U} \cdot \left(\mathbf{V}^T \cdot \mathbf{\Sigma} \right)^T = \mathbf{U} \cdot \mathbf{\Sigma} \cdot \mathbf{V}.$$

Further, let the number of positive singular values be $r \leq \min\{n, m\}$, i.e.

$$\sigma_1 \geq \ldots \geq \sigma_r > \sigma_{r+1} = \ldots = \sigma_{\min\{n,m\}} = 0.$$

Hence, we can reduce the singular value decomposition:

$$R = \mathbf{U} \cdot \mathbf{\Sigma} \cdot \mathbf{V} = (U, \star) \cdot \begin{pmatrix} \Sigma & 0 \\ 0 & 0 \end{pmatrix} \cdot \begin{pmatrix} V \\ \star \end{pmatrix} = U \cdot \Sigma \cdot V.$$

The latter provides $(\mathcal{SVD})$ in reduced form.

3.2.2.4 Rank and Positive Singular Values

It remains to relate the rank of the matrix R to the number r of its positive singular values. This will be done by showing:

$$r = \text{rank}(R).$$

Let us first consider the case $m \leq n$. Recall that the rank of the matrix R is the maximal number of its linearly independent columns. Thus, it coincides with the dimension of the *range* of R, i.e.

$$\text{rank}(R) = \dim(\text{range}(R)),$$

where

$$\text{range}(R) = \left\{ R \cdot v \mid v \in \mathbb{R}^m \right\}.$$

In particular, by considering first r left- and right-singular vectors with $\sigma_i > 0$, we have:

$$R \cdot \left(\frac{1}{\sigma_i} \cdot v_i \right) = u_i \quad \text{for all } i = 1, \dots, r.$$

The range of R contains at least r linearly independent vectors u_i, i.e.

$$\dim(\text{range}(R)) \geq r.$$

By considering last $m - r$ left- and right-singular vectors with $\sigma_i = 0$, we have:

$$R \cdot v_i = 0 \quad \text{for all } i = r + 1, \dots, m.$$

In terms of the *nullspace* of R, the latter means:

$$v_i \in \text{null}(R) = \left\{ v \in \mathbb{R}^m \mid R \cdot v = 0 \right\} \quad \text{for all } i = r + 1, \dots, m.$$

The nullspace of R contains at least $m - r$ linearly independent vectors v_i, i.e.

$$\dim(\text{null}(R)) \geq m - r.$$

Overall, we are done by additionally using the well-known identity, see e.g. Lancaster (1969):

$$\dim(\text{range}(R)) + \dim(\text{null}(R)) = m.$$

The case $m > n$ can be treated analogously by focusing on the matrix R^T, and recalling that the rank is invariant under transposition, i.e.

$$\text{rank}(R) = \text{rank}\left(R^T \right).$$

3.2.2.5 Low-Rank Approximation

We turn our attention to the crucial question on how to predict possibly missing ratings of the matrix $R = (r_{ij})$. To do so, we construct an approximation $(n \times m)$-matrix $A = (a_{ij})$. The missing ratings of R will be predicted by the corresponding entries of A. The specified ratings of R should be approximated by the corresponding entries of A as good as possible. The sum of their squared differences measures the quality of such an approximation:

$$\sqrt{\sum_{(i,j)\in S} \left(r_{ij} - a_{ij}\right)^2},$$

where the user-movie pairs of specified ratings are stored in the set

$$S = \left\{(i, j) \mid r_{ij} \text{ is specified}\right\}.$$

Moreover, we fix the rank of an approximation matrix A by requiring:

$$\text{rank}(A) = s.$$

This imposes a certain pattern on the users' choice behavior. Namely, users are assumed to rate movies by focusing on their s latent features. Overall, we are interested in *low-rank approximation* of R:

$$\min_{A=(a_{ij})} \sqrt{\sum_{(i,j)\in S} \left(r_{ij} - a_{ij}\right)^2} \quad \text{s.t.} \quad \text{rank}(A) = s. \tag{$\mathcal{A}$}$$

In what follows, we justify the low-rank approximation approach, and discuss how to efficiently solve the optimization problem $(\mathcal{A})$.

3.2.2.6 Frobenius Norm

As the main tool for our analysis of $(\mathcal{A})$, we use an appropriate matrix norm. The *Frobenius norm* of an $(n \times m)$-matrix $B = \left(b_{ij}\right)$ is defined as the square root of the sum taken over all its squared entries:

$$\|B\|_F = \sqrt{\sum_{j=1}^{m}\sum_{i=1}^{n} b_{ij}^2}.$$

In order to rewrite the latter, we compute the j-th diagonal entry of the $(m \times m)$-matrix $B^T \cdot B$:

$$\left(B^T \cdot B\right)_{jj} = \left(b_{1j}, \ldots, b_{nj}\right) \cdot \begin{pmatrix} b_{1j} \\ \vdots \\ b_{nj} \end{pmatrix} = \sum_{i=1}^{n} b_{ij}^2.$$

Summing up over $j = 1, \ldots, m$, we get the trace of $B^T \cdot B$:

$$\text{trace}\left(B^T \cdot B\right) = \sum_{j=1}^{m} \left(B^T \cdot B\right)_{jj} = \sum_{j=1}^{m} \sum_{i=1}^{n} b_{ij}^2.$$

Recall that the *trace* of a quadratic matrix is defined as the sum of its diagonal entries. Analogously, we have for the trace of the $(n \times n)$-matrix $B \cdot B^T$:

$$\text{trace}\left(B \cdot B^T\right) = \sum_{i=1}^{n} \left(B \cdot B^T\right)_{ii} = \sum_{i=1}^{n} \sum_{j=1}^{m} b_{ij}^2.$$

Comparing both formulae with the definition of the Frobenius norm, we obtain:

$$\|B\|_F = \sqrt{\text{trace}\left(B^T \cdot B\right)} = \sqrt{\text{trace}\left(B \cdot B^T\right)}.$$

From the latter, we deduce that the Frobenius norm is invariant under orthogonal transformations:

$$\|U \cdot B \cdot V\|_F^2 = \text{trace}\left((U \cdot B \cdot V)^T \cdot U \cdot B \cdot V\right) = \text{trace}\left(V^T \cdot B^T \cdot \underbrace{U^T \cdot U}_{=I} \cdot B \cdot V\right)$$

$$= \text{trace}\left((B \cdot V)^T \cdot B \cdot V\right) = \text{trace}\left(B \cdot V \cdot (B \cdot V)^T\right)$$

$$= \text{trace}\left(B \cdot \underbrace{V \cdot V^T}_{=I} B^T\right) = \text{trace}\left(B \cdot B^T\right) = \|B\|_F^2.$$

3.2.2.7 Eckart-Young-Mirsky Theorem

Now, we are ready to justify the low-rank approximation approach. For that, we assume for a moment that all entries of the rating $(n \times m)$-matrix R are specified, i.e.

$$S = \{(i, j) \mid i = 1, \ldots, n \text{ and } j = 1, \ldots, m\}.$$

Then, the optimization problem $(\mathcal{A})$ can be equivalently written in terms of the Frobenius norm:

$$\min_{A} \|R - A\|_F^2 \quad \text{s.t.} \quad \text{rank}(A) = s. \qquad (\mathcal{F})$$

It turns out that a solution of $(\mathcal{F})$ can be easily constructed by using the singular value decomposition of R. According to $(\mathcal{SVD})$, we have:

$$R = U \cdot \Sigma \cdot V,$$

where the columns of U and the rows of V are pairwise orthogonal, respectively, i.e.

$$U^T \cdot U = I, \quad V \cdot V^T = I,$$

and Σ stores on its diagonal the positive singular values of R, i.e.

$$\Sigma = \mathrm{diag}\,(\sigma_1, \ldots, \sigma_s, \sigma_{s+1}, \ldots, \sigma_r)\,.$$

Recall that we agreed to write the positive singular values of R in nondecreasing order:

$$\sigma_1 \geq \ldots \geq \sigma_s \geq \sigma_{s+1} \geq \ldots \geq \sigma_r > 0.$$

Eckart-Young-Mirsky theorem provides a best s-rank approximation of R, see e.g. Lancaster (1969):

$$A = U \cdot \Sigma_s \cdot V,$$

where in Σ_s the smallest $r - s$ singular values are set to zero, i.e.

$$\Sigma_s = \mathrm{diag}\,(\sigma_1, \ldots, \sigma_s, 0, \ldots, 0)\,.$$

First, note that the approximation matrix A is feasible for $(\mathcal{F})$. As we have seen above, the rank of a matrix corresponds to the number of its positive singular values. Since the positive singular values of A are $\sigma_1, \ldots, \sigma_s$, this observation, applied to A, ensures that

$$\mathrm{rank}(A) = s.$$

Further, we compute the error between the rating matrix R and its s-rank approximation A. By using the invariance of Frobenius norm under orthogonal transformations, we get:

$$\|R - A\|_F = \|U \cdot \Sigma \cdot V - U \cdot \Sigma_s \cdot V\|_F = \|U \cdot (\Sigma - \Sigma_s) \cdot V\|_F^2 = \|\Sigma - \Sigma_s\|_F$$

$$= \|\mathrm{diag}\,(0, \ldots, 0, \sigma_{s+1}, \ldots, \sigma_r)\|_F = \sqrt{\sigma_{s+1}^2 + \ldots + \sigma_r^2}.$$

The approximation error, thus, amounts to the Euclidean norm of $r - s$ smallest singular values of R. This justifies the low-approximation approach. In case that the rating matrix

R is completely specified, we capture s most important of r available features, which determine the underlying users' choice behavior. By solving $(\mathcal{F})$, the information on just $r - s$ dispensable features is lost.

3.2.2.8 Matrix Factorization

Finally, we turn our attention to the question on how a low-rank approximation of the rating $(n \times m)$-matrix R can be efficiently found. Hereby, we assume that some entries of R are missing. The main obstacle for developing optimization methods for $(\mathcal{A})$ is the rank feasibility condition:

$$\text{rank}(A) = s.$$

Indeed, while trying to decrease the objective function of $(\mathcal{A})$, we need to ensure that the rank of the approximation remains s. To avoid this challenging verification, we apply *matrix factorization* technique. We look for an s-rank approximation which can be represented as a product of two matrices:

$$A = X \cdot Y,$$

where $X = (x_{ik})$ is the left factor $(n \times s)$-matrix and $Y = (y_{kj})$ is the right factor $(s \times m)$-matrix. Generically, the left and right factor matrices are of full rank, i.e.

$$\text{rank}(X) = \text{rank}(Y) = s.$$

Loosely speaking, the genericity means here that, if we choose the entries of X and Y randomly and independently from each other, all s columns of X and all s rows of Y will be linearly independent with probability one. It remains to note that in this case the rank of their matrix product $A = X \cdot Y$ is automatically s, see Exercise 3.5. By applying the matrix factorization, we have a substitute for the entries of the low-rank approximation A:

$$a_{ij} = \sum_{k=1}^{s} x_{ik} \cdot y_{kj}.$$

Then, the relaxation of the optimization problem $(\mathcal{A})$ reads in new variables as follows:

$$\min_{\substack{X=(x_{ik}) \\ Y=(y_{kj})}} \sqrt{\sum_{(i,j) \in S} \left(r_{ij} - \sum_{k=1}^{s} x_{ik} \cdot y_{kj} \right)^2}. \tag{$\mathcal{R}$}$$

Why is it more advantageous to consider $(\mathcal{R})$ instead of $(\mathcal{A})$? First, the absence of the hard rank constraint facilitates the numerical treatment. Another advantage comes from the fact that the number of features is much less than the number of users and movies, i.e.

$$s \ll \min\{n, m\}.$$

Let us then compare the numbers of variables in ($\mathcal{A}$) and ($\mathcal{R}$).

$$\#\text{variables}(\mathcal{A}) = \underbrace{n \cdot m}_{\text{size of A}} \ , \quad \#\text{variables}(\mathcal{R}) = \underbrace{n \cdot s}_{\text{size of X}} + \underbrace{s \cdot m}_{\text{size of Y}} = s \cdot (n + m).$$

Hence, by switching from ($\mathcal{A}$) to ($\mathcal{R}$), we considerably reduced the number of variables, which keeps the necessary computational efforts bounded.

3.2.2.9 Gradient Descent

We intend to iteratively solve the relaxed optimization problem ($\mathcal{R}$) by an application of gradient descent. Let us for convenience square and scale its objective function:

$$f(X, Y) = \frac{1}{2} \cdot \sum_{(i,j) \in S} \left(r_{ij} - \sum_{k=1}^{s} x_{ik} \cdot y_{kj} \right)^2.$$

How is it possible to approach the minimizers of f? To find a local minimum of a function using gradient descent, we take steps proportional to the negative of its gradient at the current iterate. The reason is that the function decreases fastest in the gradient direction. In matrix form, the iteration scheme of gradient descent with fixed step-size $\eta > 0$ reads as follows:

$$\begin{array}{c} \text{iteration} \\ \text{step} \end{array} \left\{ \begin{array}{l} X(t+1) - X(t) = -\eta \cdot \nabla_X f(X(t), Y(t)) \\[2mm] Y(t+1) - Y(t) = -\eta \cdot \nabla_Y f(X(t), Y(t)) \end{array} \right\} \begin{array}{c} \text{gradient} \\ \text{direction} \end{array}$$

Here, the difference of the next $(t + 1)$-th and the previous t-th iterate is taken to be proportional to the gradient of f. Aiming to make our iteration scheme explicit, we first compute the partial derivatives:

$$\frac{\partial f(X, Y)}{\partial x_{ik}} = - \sum_{j:(i,j) \in S} \left(r_{ij} - \sum_{k=1}^{s} x_{ik} \cdot y_{kj} \right) \cdot y_{kj},$$

$$\frac{\partial f(X, Y)}{\partial y_{kj}} = - \sum_{i:(i,j) \in S} \left(r_{ij} - \sum_{k=1}^{s} x_{ik} \cdot y_{kj} \right) \cdot x_{ik}.$$

Second, we rewrite these formulae by using the error $(n \times m)$-matrix $E = (e_{ij})$ with entries

$$e_{ij} = \begin{cases} r_{ij} - \displaystyle\sum_{k=1}^{s} x_{ik} \cdot y_{kj}, & \text{if } r_{ij} \text{ is specified,} \\[2em] 0, & \text{if } r_{ij} \text{ is missing.} \end{cases}$$

Substituting into partial derivatives of f, we get in matrix form:

$$\nabla_X f(X, Y) = -E \cdot Y^T, \quad \nabla_Y f(X, Y) = -X^T \cdot E.$$

Finally, we obtain the *gradient descent* for solving the relaxed optimization problem $(\mathcal{R})$:

$$X(t+1) = X(t) + \eta \cdot E(t) \cdot Y^T(t),$$

$$Y(t+1) = Y(t) + \eta \cdot X^T(t) \cdot E(t). \tag{$\mathcal{GD}$}$$

Let us apply $(\mathcal{GD})$ to the Netflix matrix from above. We assume that the users' choice behavior is guided by two latent features, i.e. $s = 2$. After some iterations the model-based completion is roughly as follows, cf. Exercise 3.7:

$$R_{model} = \begin{pmatrix} & \boxed{M1} & \boxed{M2} & \boxed{M3} & \boxed{M4} \\ \boxed{U1} & 4.99 & 2.97 & \mathbf{1.25} & 0.94 \\ \boxed{U2} & 3.97 & \mathbf{2.38} & 1.24 & 0.96 \\ \boxed{U3} & 0.96 & 1.01 & \mathbf{5.54} & 4.81 \\ \boxed{U4} & 0.99 & \mathbf{0.93} & 4.44 & 3.85 \\ \boxed{U5} & \mathbf{0.97} & 0.95 & \mathbf{4.76} & 4.13 \end{pmatrix}.$$

We see that the specified ratings of R do not change much within R_{model}. Together with completed missing ratings they reveal the hidden pattern of users' choice behavior. The positive singular values of the completion R_{model} can be computed numerically:

$$\sigma_1 = 12.36, \quad \sigma_2 = 6.41, \quad \sigma_3 = 0.0048, \quad \sigma_4 = 0.0018.$$

This is in accordance with our assumption that just two features are significant. The third and the fourth feature can be neglected, since with $\sigma_3 = 0.0048$ and $\sigma_4 = 0.0018$ they are of relatively small importance. Further, note that the model- and neighborhood-based completions considerably differ from each other. In particular, we recommend for the user U2 to see the movie M2 due to the model-based completion R_{model}. This is in strong contrast to the recommendation from the neighborhood-based completion $R_{neighbor}$. There, we recommended for the user U2 to see rather the movie M3.

3.3   Case Study: Latent Semantic Analysis

Latent semantic analysis is a technique in natural language processing of analyzing relationships between the terms and documents containing them. This is done by identifying some intermediate concepts related to the terms and documents. Latent semantic analysis assumes the *distributional hypothesis*: terms with similar distributions have similar meanings. Or as it has been put by Firth (1957), a word is characterized by the company it keeps. The general idea behind the distributional hypothesis seems rather reasonable: there is a correlation between distributional similarity and meaning similarity, which allows us to utilize the former in order to estimate the latter, cf. Sahlgren (2008).

Let us become acquainted with the latent semantic analysis on an example. Imagine that a holiday portal hires you, in order to improve travel suggestions for their users. You have access to a database, which contains a bunch of user reviews on the destinations D1–D6. The relevant document extracts are listed below:

D1: I spent one week at **Lake** Garda, which is beautifully surrounded by **hills**. There were a lot of sport and wellness offers I tried, but most of the time I relaxed at the **beach** and enjoyed the sun.

D2: We had astonishing **beach** holidays at the **sea**. Weather and hotel have been perfect.

D3: The only things I want to do on holidays is relaxing at the water, while catching some sun. Our balcony provided a nice view of the **lake** with some **mountains** behind.

D4: We had a short hiking trip to the **hills** and **mountains** of Austria.

D5: We made a bike tour through the **hills**, because we like the quiet environment. Fortunately, we had also time for a day in the beautiful spa area.

D6: I hate **beaches**, but love **mountains** and snow. The slopes were perfectly prepared for skiing. We had a great day!

Task 1 Build a *term-document matrix* $F = (f_{ij})$ out of the reviews, where the relevant terms T1–T5 "lake", "sea", "beach", "hills", and "mountains" are marked in bold. Its entry f_{ij} corresponds to the binary frequency, i.e. it equals 1 if the term i occurs in the document j, and 0 otherwise. Interpret the columns $x_1, \ldots, x_6 \in \{0, 1\}^5$ of F as the representations of the documents D1–D6 with respect to the terms T1–T5.

Hint 1 The term-document matrix is given as follows:

$$
F = \begin{pmatrix}
 & D1 & D2 & D3 & D4 & D5 & D6 \\
T1 & 1 & 0 & 1 & 0 & 0 & 0 \\
T2 & 0 & 1 & 0 & 0 & 0 & 0 \\
T3 & 1 & 1 & 0 & 0 & 0 & 1 \\
T4 & 1 & 0 & 0 & 1 & 1 & 0 \\
T5 & 0 & 0 & 1 & 1 & 0 & 1
\end{pmatrix}.
$$

Task 2 A potential user puts the following text query:

Q: Holiday at the sea with mountain biking opportunities.

Represent the query Q as a vector $q \in \{0, 1\}^5$ of binary frequencies with respect to the terms T1–T5. Compute the cosine similarity between the representations q of the query Q and $x_1, \ldots, x_6$ of the documents D1–D6, respectively. Based on that, which destination should be suggested to the user? Is the resulted recommendation satisfactory?

Hint 2 With $q = (0, 1, 0, 0, 1)^T$, the cosine similarities are easily calculated:

$$\text{Cosine}\,(q, x_1) = 0, \quad \text{Cosine}\,(q, x_2) = \tfrac{1}{2}, \quad \text{Cosine}\,(q, x_3) = \tfrac{1}{2},$$

$$\text{Cosine}\,(q, x_4) = \tfrac{1}{2}, \quad \text{Cosine}\,(q, x_5) = 0, \quad \text{Cosine}\,(q, x_6) = \tfrac{1}{2}.$$

The most similar destinations are D2, D3, D4, and D6. It is not clear which one to choose.

Task 3 Compute a best 2-rank approximation A of the term-document matrix F. Find the singular-value decomposition of A in reduced form, i.e.

$$A = U \cdot \Sigma_2 \cdot V,$$

where U is a term-concept (5×2)-matrix with orthogonal columns, V is a concept-document (2×6)-matrix with orthogonal rows, and Σ is the diagonal (2×2)-matrix with the singular values of A placed on the main diagonal in nondecreasing order.

Hint 3 A best 2-rank approximation of F is $A = U \cdot \Sigma_2 \cdot V$ with

$$
U = \begin{pmatrix} & \text{C1} & \text{C2} \\ \text{T1} & 0.40 & 0.12 \\ \text{T2} & 0.11 & -0.48 \\ \text{T3} & 0.54 & -0.69 \\ \text{T4} & 0.51 & 0.37 \\ \text{T5} & 0.51 & 0.37 \end{pmatrix}, \quad
V = \begin{pmatrix} & \text{D1} & \text{D2} & \text{D3} & \text{D4} & \text{D5} & \text{D6} \\ \text{C1} & 0.61 & 0.27 & 0.38 & 0.42 & 0.21 & 0.44 \\ \text{C2} & -0.13 & -0.75 & 0.31 & 0.48 & 0.24 & -0.20 \end{pmatrix},
$$

$$\Sigma_2 = \text{diag}(2.42, 1.56).$$

Task 4 Interpret the columns of V as 2-dimensional representations $\tilde{x}_1, \ldots, \tilde{x}_6$ of the documents D1–D6 with respect to the concepts C1–C2. By setting $q = U \cdot \Sigma_2 \cdot \tilde{q}$, find for the query Q an analogous 2-dimensional representation $\tilde{q}$.

Hint 4 We deduce:

$$\tilde{q} = \underbrace{\Sigma_2^{-1} \cdot \Sigma_2}_{=I} \cdot \tilde{q} = \Sigma_2^{-1} \cdot \underbrace{U^T \cdot U}_{=I} \cdot \Sigma_2 \cdot \tilde{q} = \Sigma_2^{-1} \cdot U^T \cdot \underbrace{U \cdot \Sigma_2 \cdot \tilde{q}}_{=q} = \Sigma_2^{-1} \cdot U^T \cdot q.$$

Finally, it holds:

$$\tilde{q} = \Sigma_2^{-1} \cdot U^T \cdot q = \begin{pmatrix} 2.42 & 0 \\ 0 & 1.56 \end{pmatrix}^{-1} \cdot \begin{pmatrix} 0.40 & 0.12 \\ 0.11 & -0.48 \\ 0.54 & -0.69 \\ 0.51 & 0.37 \\ 0.51 & 0.37 \end{pmatrix}^T \cdot \begin{pmatrix} 0 \\ 1 \\ 0 \\ 0 \\ 1 \end{pmatrix} = \begin{pmatrix} 0.26 \\ -0.07 \end{pmatrix}.$$

Task 5 Compute the cosine similarity between the 2-dimensional representations $\tilde{q}$ of the query Q and $\tilde{x}_1, \ldots, \tilde{x}_6$ of the documents D1–D6, respectively. Based on that, which destination should be suggested to the user? Is now the resulted recommendation satisfactory?

Hint 5 The cosine similarities are easily calculated:

$$\text{Cosine } (\tilde{q}, \tilde{x}_1) = 1.00, \quad \text{Cosine } (\tilde{q}, \tilde{x}_2) = 0.57, \quad \text{Cosine } (\tilde{q}, \tilde{x}_3) = 0.58,$$

$$\text{Cosine } (\tilde{q}, \tilde{x}_4) = 0.44, \quad \text{Cosine } (\tilde{q}, \tilde{x}_5) = 0.44, \quad \text{Cosine } (\tilde{q}, \tilde{x}_6) = 0.99.$$

Surprisingly enough, the destination D1 should be recommended. Though the document D1 and the query Q do not have common terms, their descriptions both contain the relevant concepts of "water" and "hiking". Due to the same reason, the document D6 has also a high similarity to the query Q. Note that our choice of terms neglects the negative attitude towards the beaches in the document D6.

3.4   Exercises

Exercise 3.1 (User and Movie Similarity) Given is a Netflix rating matrix:

$$R = \begin{pmatrix} & \text{M1} & \text{M2} & \text{M3} & \text{M4} \\ \text{U1} & 5 & 3 & - & 1 \\ \text{U2} & 4 & - & - & 1 \\ \text{U3} & 1 & 1 & - & 5 \\ \text{U4} & 1 & - & - & 4 \\ \text{U5} & - & 1 & 5 & 4 \end{pmatrix}.$$

Complete R by applying the $(k\mathcal{N}\mathcal{N})$ algorithm. Take the cosine similarity for users, on one hand, and for movies, on the other hand. Make comparisons between the two cases.

Exercise 3.2 (Eigenvalues and Singular Values) Let an $(n \times m)$-matrix R with positive singular values $\sigma_1, \ldots, \sigma_r$ be given. Show the following assertions:

(i) There are exactly r positive eigenvalues of $R^T \cdot R$ and $R \cdot R^T$, respectively.
(ii) The positive singular values of R coincide with the square roots of the positive eigenvalues $\lambda_1, \ldots, \lambda_r$ of $R^T \cdot R$ and $R \cdot R^T$, i.e.

$$\sigma_i = \sqrt{\lambda_i} \quad \text{for all } i = 1, \ldots, r.$$

Exercise 3.3 (Frobenius Norm and Singular Values) Let an $(n \times m)$-matrix R with positive singular values $\sigma_1, \ldots, \sigma_r$ be given. Show that the Frobenius norm of R can be represented by means of its singular values:

$$\|R\|_F = \sqrt{\sigma_1^2 + \ldots + \sigma_r^2}.$$

Exercise 3.4 (Largest and Smallest Singular Values) Show that for the largest and smallest singular values of an $(n \times m)$-matrix R holds:

$$\sigma_{max}(R) = \max_{\|z\|_2=1} \|R \cdot z\|_2, \quad \sigma_{min}(R) = \min_{\|z\|_2=1} \|R \cdot z\|_2.$$

Exercise 3.5 (Features) Let the following user-movie rating matrix be given:

$$R = \begin{pmatrix} 1 & 2 & 4 & -9 \\ 1 & 1 & 3 & -6 \\ -5 & 6 & -4 & -3 \end{pmatrix}.$$

(i) Compute the number of latent features and their importance for all users collectively.
(ii) Find a best 1-rank approximation of R and compute the corresponding error.

Exercise 3.6 (Rank of Matrix Product) Let an $(n \times s)$-matrix X and an $(s \times m)$-matrix Y be given. Assume that X and Y are of full rank $s \le \min\{n, m\}$. Show that the rank of their matrix product $A = X \cdot Y$ also equals to s.

Exercise 3.7 (Low-Rank Approximation) Complete the Netflix matrix R from Exercise 3.1 by computing its 2-rank approximation. Apply $(\mathcal{G}\mathcal{D})$ for the latter computation.

Classification

4

Classification is a process by which new objects, events, people, or experiences are assigned to some class on the basis of characteristics shared by members of the same class, and features distinguishing the members of one class from those of another. In the context of data science it is often necessary to categorize new, unlabeled information based upon its relevance to known, labeled data. Usual applications include *credit investigation* of a potential client in presence of the current or previous clients with disclosed financial history. Another important application deals with the analytical *quality control*. Here, a decision has to be made whether a patient is likely to be virus-infected by comparing own and other patients' test results. In this chapter, we shall use *linear classifiers* to assign a newcomer to a particular class. This assignment depends on whether the corresponding performance of the newcomer exceeds a certain bound. Three types of linear classifiers are discussed. First, we introduce the statistically motivated *Fisher's discriminant*. The latter maximizes the sample variance between the classes and minimizes the variance of data within the classes. The computation of Fisher's discriminant leads to a nicely structured eigenvalue problem. Second, the celebrated *support-vector machine* is studied. It is geometrically motivated, and maximizes the margin between two classes. The detection of an optimal separating hyperplane is based on the convex duality. Third, the *naïve Bayes classifier* is derived. Rooted in the application of Bayes theorem, the latter is of probabilistic origin. Namely, the Bernoulli probabilities of an assignment to one or another class conditioned on the observed data are compared.

4.1 Motivation: Credit Investigation

A *credit investigation* is a procedure undertaken by a financial institution to vet a potential client's ability to pay back a loan. Failure to pass this procedure means disapproval of a loan. When clients apply for a loan from a bank, they are asked to disclose details

© Springer-Verlag GmbH Germany, part of Springer Nature 2021
V. Shikhman, D. Müller, *Mathematical Foundations of Big Data Analytics*,
https://doi.org/10.1007/978-3-662-62521-7_4

about their financial history. Based on this information, a forecast on the newcomer's creditworthiness is produced. Thus, a potential client has to be classified into one of two classes depending on the expectation of repaying a loan. This classification is based on the data set of current or previous clients who disclosed their financial history, were granted a loan, and turned out to be creditworthy or not.

Let us mathematically model the problem of credit investigation. For that, n clients with known profiles on their financial situation are given:

$$x_1, \ldots, x_n \in \mathbb{R}^m,$$

where m is the number of relevant *features*, e.g. income, debts, salary etc. Due to their disclosed financial history, we subdivide the clients into the classes C_{yes} and C_{no} depending on whether they have payed back a loan or not. Equivalently, the clients are labeled by $y_i \in \{\pm 1\}$, $i = 1, \ldots, n$, according to their *creditworthiness*:

$$y_i = \begin{cases} +1, \text{ if } i \in C_{yes}, \\ -1, \text{ if } i \in C_{no}. \end{cases}$$

Here, $y_i = +1$ means that the i-th client has been credible in the past, and $y_i = -1$ means the opposite. The problem of credit investigation is to meaningfully forecast if a new client with the disclosed profile $x \in \mathbb{R}^m$ is rather creditworthy or not. In other words, we first need to compare the newcomer's profile x with those $x_1, \ldots, x_n$ within the provided data set. Based on this comparison, we shall be hopefully able to classify the newcomer, i.e. to make an assignment to the class C_{yes} or C_{no}. Thus, the newcomer is to be labeled by either $y = +1$ or $y = -1$.

In this chapter, we shall approach the classification task by means of linear classifiers. A *linear classifier* $a \in \mathbb{R}^m$ is a vector of weights which correspond to the features. The performance of the feature vector $x \in \mathbb{R}^m$ is then the scalar product $a^T \cdot x$. In order to classify x, we shall be guided by its performance $a^T \cdot x$, rather than by the feature vector x itself. More precisely, if the performance $a^T \cdot x$ is above a certain *bound* $b \in \mathbb{R}$, the newcomer x is assigned to the class C_{yes}, and if it is below to the class C_{no}. The *linear classifier rule* reads:

$$y = \begin{cases} +1, \text{ if } a^T \cdot x \geq b, \\ -1, \text{ else.} \end{cases} \tag{$\mathcal{LC}$}$$

Now, the crucial question is on how to properly choose the linear classifier a and the bound b in dependence on the data set $(x_i, y_i) \in \mathbb{R}^m \times \{\pm 1\}$, $i = 1, \ldots, n$.

4.2   Results

4.2.1   Fisher's Discriminant Rule

Let us present the Fisher's discriminant rule for classification, see e.g. Mathar et al. (2020). Its motivation is rooted mainly in statistical considerations, involving sample mean and sample variance.

4.2.1.1 Sample Mean

The *Fisher's discriminant rule* compares a newcomer x with the classes' *sample means*:

$$x_{yes} = \frac{1}{n_{yes}} \cdot \sum_{i \in C_{yes}} x_i, \quad x_{no} = \frac{1}{n_{no}} \cdot \sum_{i \in C_{no}} x_i,$$

where n_{yes} and n_{no} denote the numbers of data points in C_{yes} and C_{no}, respectively. It is reasonable to label x by $+1$ or -1 depending on whether its performance $a^T \cdot x$ is closer to the average performance $a^T \cdot x_{yes}$ of the class C_{yes} or $a^T \cdot x_{no}$ of the class C_{no}, i.e.

$$y = \begin{cases} +1, \text{ if } \left| a^T \cdot x - a^T \cdot x_{yes} \right| \leq \left| a^T \cdot x - a^T \cdot x_{no} \right|, \\[2mm] -1, \text{ else.} \end{cases} \qquad (\mathcal{FD})$$

The choice of the vector $a \in \mathbb{R}^m$ will be statistically motivated. Loosely speaking, the average performances of the classes should differ the most, and the performances of the data points within the classes should differ the least. With respect to their performance, the classes should be far away from each other, but at the same time their data points should be relatively close to each other, see Fig. 4.1. We capture this intuition by the notion of sample variance.

4.2.1.2 Sample Variance

Let us consider the *sample variance* of the data points with respect to their performance:

$$\text{Var}(a) = \frac{1}{n} \cdot \sum_{i=1}^{n} \left(a^T \cdot x_i - a^T \cdot \bar{x} \right)^2,$$

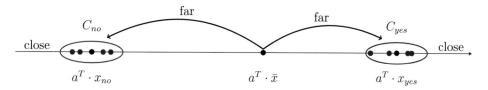

Fig. 4.1 Fisher discriminant

where the sample mean of data points is given by

$$\bar{x} = \frac{1}{n} \cdot \sum_{i=1}^{n} x_i.$$

In what follows, we decompose the sample variance $\mathrm{Var}(a)$ into two parts:

- the within-classes variance $W(a)$,
- the between-classes variance $B(a)$.

In order to simplify the notation, we proceed by setting:

$$z_i = a^T \cdot x_i, \quad z_{yes} = a^T \cdot x_{yes}, \quad z_{no} = a^T \cdot x_{no}, \quad \bar{z} = a^T \cdot \bar{x}.$$

It holds:

$$\mathrm{Var}(a) = \frac{1}{n} \cdot \sum_{i=1}^{n} \left(a^T \cdot x_i - a^T \cdot \bar{x} \right)^2 = \frac{1}{n} \cdot \sum_{i=1}^{n} (z_i - \bar{z})^2$$

$$= \frac{1}{n} \cdot \left(\sum_{i \in C_{yes}} (z_i - z_{yes} + z_{yes} - \bar{z})^2 + \sum_{i \in C_{no}} (z_i - z_{no} + z_{no} - \bar{z})^2 \right)$$

$$= \frac{1}{n} \cdot \left(\sum_{i \in C_{yes}} \left((z_i - z_{yes})^2 + (z_{yes} - \bar{z})^2 \right) + \sum_{i \in C_{no}} \left((z_i - z_{no})^2 + (z_{no} - \bar{z})^2 \right) \right)$$

$$+ \frac{1}{n} \cdot \left(\sum_{i \in C_{yes}} 2 (z_i - z_{yes}) \cdot (z_{yes} - \bar{z}) + \sum_{i \in C_{no}} 2 (z_i - z_{no}) \cdot (z_{no} - \bar{z}) \right).$$

It is not hard to see that the last term vanishes, see Exercise 4.2:

$$\sum_{i \in C_{yes}} (z_i - z_{yes}) \cdot (z_{yes} - \bar{z}) + \sum_{i \in C_{no}} (z_i - z_{no}) \cdot (z_{no} - \bar{z}) = 0.$$

Hence, we get by rearranging:

$$\mathrm{Var}(a) = W(a) + B(a),$$

where the within-classes variance is given by

$$W(a) = \frac{1}{n} \cdot \left(\sum_{i \in C_{yes}} (z_i - z_{yes})^2 + \sum_{i \in C_{no}} (z_i - z_{no})^2 \right)$$

$$= \underbrace{\frac{n_{yes}}{n}}_{\substack{\text{weight} \\ \text{of } C_{yes}}} \cdot \underbrace{\frac{1}{n_{yes}} \cdot \sum_{i \in C_{yes}} (z_i - z_{yes})^2}_{\substack{\text{variance} \\ \text{within } C_{yes}}} + \underbrace{\frac{n_{no}}{n}}_{\substack{\text{weight} \\ \text{of } C_{no}}} \cdot \underbrace{\frac{1}{n_{no}} \cdot \sum_{i \in C_{no}} (z_i - z_{no})^2}_{\substack{\text{variance} \\ \text{within } C_{no}}},$$

and the between-classes variance by

$$B(a) = \frac{1}{n} \cdot \left(\sum_{i \in C_{yes}} (z_{yes} - \bar{z})^2 + \sum_{i \in C_{no}} (z_{no} - \bar{z})^2 \right)$$

$$= \underbrace{\frac{n_{yes}}{n}}_{\substack{\text{weight} \\ \text{of } C_{yes}}} \cdot \underbrace{(z_{yes} - \bar{z})^2}_{\substack{\text{deviation} \\ \text{of } C_{yes}}} + \underbrace{\frac{n_{no}}{n}}_{\substack{\text{weight} \\ \text{of } C_{no}}} \cdot \underbrace{(z_{no} - \bar{z})^2}_{\substack{\text{deviation} \\ \text{of } C_{no}}}.$$

By substituting the vector a back, we have for the squares above:

$$(z_i - z_{yes})^2 = (a^T \cdot x_i - a^T \cdot x_{yes})^2 = a^T \cdot (x_i - x_{yes}) \cdot (x_i - x_{yes})^T \cdot a,$$

$$(z_i - z_{no})^2 = (a^T \cdot x_i - a^T \cdot x_{no})^2 = a^T \cdot (x_i - x_{no}) \cdot (x_i - x_{no})^T \cdot a,$$

and, analogously:

$$(z_{yes} - \bar{z})^2 = (a^T \cdot x_{yes} - a^T \cdot \bar{x})^2 = a^T \cdot (x_{yes} - \bar{x}) \cdot (x_{yes} - \bar{x})^T \cdot a,$$

$$(z_{no} - \bar{z})^2 = (a^T \cdot x_{no} - a^T \cdot \bar{x})^2 = a^T \cdot (x_{no} - \bar{x}) \cdot (x_{no} - \bar{x})^T \cdot a.$$

Note that we use here the so-called dyadic product of two vectors $u, w \in \mathbb{R}^m$ resulting in an $(m \times m)$-matrix $u \cdot w^T$ of rank one. Overall, the following representation of the within- and between-classes variances can be obtained:

$$W(a) = a^T \cdot W \cdot a, \quad B(a) = a^T \cdot B \cdot a$$

with the $(m \times m)$-matrices:

$$W = \frac{1}{n} \cdot \left(\sum_{i \in C_{yes}} \left(x_i - x_{yes}\right) \cdot \left(x_i - x_{yes}\right)^T + \sum_{i \in C_{no}} \left(x_i - x_{no}\right) \cdot \left(x_i - x_{no}\right)^T \right),$$

$$B = \frac{n_{yes}}{n} \cdot \left(x_{yes} - \bar{x}\right) \cdot \left(x_{yes} - \bar{x}\right)^T + \frac{n_{no}}{n} \cdot \left(x_{no} - \bar{x}\right) \cdot \left(x_{no} - \bar{x}\right)^T.$$

4.2.1.3 Fisher's Discriminant

Now, we are ready to describe how to determine the vector a in a statistically reasonable way. Namely, the between-classes variance $B(a)$ of the performances should be maximized, and the within-classes variance $W(a)$ of the performances should be minimized. This can be achieved by maximizing their ratio:

$$\max_a \ \frac{B(a)}{W(a)} = \frac{a^T \cdot B \cdot a}{a^T \cdot W \cdot a}. \tag{$\mathcal{V}$}$$

A solution a of the optimization problem $(\mathcal{V})$ is called the *Fisher's discriminant*. The latter has been introduced by Fisher (1936) in the context of plant taxonomy. Let us turn our attention to the computation of the Fisher's discriminant. First, note that the matrices W and B are symmetric and positive semidefinite as sums of dyadic vector products, i.e. for all $a \in \mathbb{R}^m$ it holds:

$$a^T \cdot W \cdot a \geq 0, \quad a^T \cdot B \cdot a \geq 0.$$

We additionally assume that the matrix W is regular. Under this assumption the optimization problem $(\mathcal{V})$ is well-defined, since the denominator is always positive. Further, we see that the numerator and denominator of the objective function in $(\mathcal{V})$ are both quadratic, or, in other words, homogeneous of degree two. Then, we may equivalently solve, see Exercise 4.3:

$$\max_a \ a^T \cdot B \cdot a \quad \text{s.t.} \quad a^T \cdot W \cdot a = 1. \tag{$\mathcal{E}$}$$

By introducing a multiplier $\mu \in \mathbb{R}$ for the equality constraint, the Lagrange multiplier rule reads as follows, see e.g. Jongen et al. (2004):

$$\nabla \left(a^T \cdot B \cdot a\right) = \mu \cdot \nabla \left(a^T \cdot W \cdot a - 1\right).$$

We obtain:

$$B \cdot a = \mu \cdot W \cdot a.$$

Since W is assumed to be regular, the Fisher's discriminant fulfills:

$$W^{-1} \cdot B \cdot a = \mu \cdot a.$$

Hence, a is an eigenvector of the matrix $W^{-1} \cdot B$. Moreover, the corresponding eigenvalue coincides with the optimal value of the optimization problem $(\mathcal{E})$:

$$\mu = \mu \cdot \underbrace{a^T \cdot W \cdot a}_{=1} = a^T \cdot \underbrace{\mu \cdot W \cdot a}_{=B \cdot a} = a^T \cdot B \cdot a.$$

Thus, we conclude that the optimization problem $(\mathcal{V})$ is solved by an *eigenvector* of the matrix $W^{-1} \cdot B$ corresponding to its largest eigenvalue.

4.2.1.4 Linear Classifier

Finally, we show that the Fisher's discriminant is a linear classifier, i.e. the Fisher's discriminant rule $(\mathcal{FD})$ fits into the framework of the linear classifier rule $(\mathcal{LC})$. For that, let us provide an explicit formula for the eigenvector of the matrix $W^{-1} \cdot B$ corresponding to its largest eigenvalue. We use $n = n_{yes} + n_{no}$ and $\bar{x} = \frac{n_{yes}}{n} \cdot x_{yes} + \frac{n_{no}}{n} \cdot x_{no}$ to derive:

$$B = \frac{n_{yes}}{n} \cdot \left(x_{yes} - \bar{x}\right) \cdot \left(x_{yes} - \bar{x}\right)^T + \frac{n_{no}}{n} \cdot (x_{no} - \bar{x}) \cdot (x_{no} - \bar{x})^T$$

$$= \frac{n_{yes}}{n} \cdot \left(x_{yes} - \left(\frac{n_{yes}}{n} \cdot x_{yes} + \frac{n_{no}}{n} \cdot x_{no}\right)\right) \cdot \left(x_{yes} - \left(\frac{n_{yes}}{n} \cdot x_{yes} + \frac{n_{no}}{n} \cdot x_{no}\right)\right)^T$$

$$+ \frac{n_{no}}{n} \cdot \left(x_{no} - \left(\frac{n_{yes}}{n} \cdot x_{yes} + \frac{n_{no}}{n} \cdot x_{no}\right)\right) \cdot \left(x_{no} - \left(\frac{n_{yes}}{n} \cdot x_{yes} + \frac{n_{no}}{n} \cdot x_{no}\right)\right)^T$$

$$= \frac{n_{yes} \cdot n_{no}^2}{n^3} \cdot \left(x_{yes} - x_{no}\right) \cdot \left(x_{yes} - x_{no}\right)^T + \frac{n_{no} \cdot n_{yes}^2}{n^3} \cdot \left(x_{no} - x_{yes}\right) \cdot \left(x_{no} - x_{yes}\right)^T$$

$$= \frac{n_{yes} \cdot n_{no} \cdot \left(n_{no} + n_{yes}\right)}{n^3} \cdot \left(x_{yes} - x_{no}\right) \cdot \left(x_{yes} - x_{no}\right)^T$$

$$= \frac{n_{yes} \cdot n_{no}}{n^2} \cdot \left(x_{yes} - x_{no}\right) \cdot \left(x_{yes} - x_{no}\right)^T.$$

We see that B can be represented as a dyadic product of the vector $x_{yes} - x_{no}$ by itself, hence, it is of rank one. The rank of the matrix $W^{-1} \cdot B$ is also one. Therefore, $W^{-1} \cdot B$ has exactly one eigenvalue different from zero. Let us show that the corresponding eigenvector is

$$a = W^{-1} \cdot \left(x_{yes} - x_{no}\right).$$

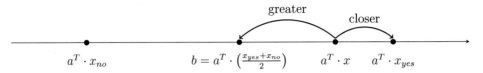

Fig. 4.2 Linear classifier

Due to the positive semidefiniteness of W, hence, also of W^{-1}, we have:

$$W^{-1} \cdot B \cdot a = W^{-1} \cdot \frac{n_{yes} \cdot n_{no}}{n^2} \cdot (x_{yes} - x_{no}) \cdot \underbrace{(x_{yes} - x_{no})^T \cdot W^{-1} \cdot (x_{yes} - x_{no})}_{\in \mathbb{R}}$$

$$= \underbrace{\frac{n_{yes} \cdot n_{no}}{n^2} \cdot (x_{yes} - x_{no})^T \cdot W^{-1} \cdot (x_{yes} - x_{no})}_{=\mu \geq 0} \cdot \underbrace{W^{-1} \cdot (x_{yes} - x_{no})}_{=a} = \mu \cdot a.$$

This justifies that a is indeed an eigenvector of $W^{-1} \cdot B$ corresponding to the largest eigenvalue μ, hence, the Fisher's discriminant.

Aiming to derive a bound b in $(\mathcal{LC})$ for the Fisher discriminant a, we first note that the average performance of the class C_{yes} is higher than that of C_{no}:

$$a^T \cdot x_{yes} - a^T \cdot x_{no} = a^T \cdot (x_{yes} - x_{no}) = \left(W^{-1} \cdot (x_{yes} - x_{no})\right)^T \cdot (x_{yes} - x_{no})$$

$$= (x_{yes} - x_{no})^T \cdot W^{-1} \cdot (x_{yes} - x_{no}) \geq 0.$$

As it can be seen from Fig. 4.2, the condition

$$\left| a^T \cdot x - a^T \cdot x_{yes} \right| \leq \left| a^T \cdot x - a^T \cdot x_{no} \right|$$

in $(\mathcal{FD})$ for labeling x by $+1$, is equivalent to

$$a^T \cdot x \geq a^T \cdot \left(\frac{x_{yes} + x_{no}}{2} \right) = b.$$

Overall, the Fisher's discriminant $a = W^{-1} \cdot (x_{yes} - x_{no})$ is a linear classifier with the bound:

$$b = \frac{(x_{yes} - x_{no})^T \cdot W^{-1} \cdot (x_{yes} + x_{no})}{2}.$$

4.2.1.5 Maximum Likelihood Classifier
Let us relate the Fisher's discriminant rule to the *likelihood maximization*. For that, we assume that the distributions of the classes C_{yes} and C_{no} are multivariate *Gaussian*

$\mathcal{N}\left(\mu_{yes}, \Sigma_{yes}\right)$ and $\mathcal{N}\left(\mu_{no}, \Sigma_{no}\right)$, where $\mu_{yes}, \mu_{no} \in \mathbb{R}^m$ are means, and $\Sigma_{yes}, \Sigma_{no} \in \mathbb{R}^{m \times m}$ are covariance matrices, respectively. The conditional probability densities of both classes are

$$p\left(x \mid C_{yes}\right) = \frac{1}{(2\pi)^{m/2} \cdot \sqrt{\det\left(\Sigma_{yes}\right)}} \cdot e^{-\frac{1}{2}(x-\mu_{yes})^T \cdot \Sigma_{yes}^{-1} \cdot (x-\mu_{yes})},$$

$$p\left(x \mid C_{no}\right) = \frac{1}{(2\pi)^{m/2} \cdot \sqrt{\det\left(\Sigma_{no}\right)}} \cdot e^{-\frac{1}{2}(x-\mu_{no})^T \cdot \Sigma_{no}^{-1} \cdot (x-\mu_{no})}.$$

By applying *Bayes theorem*, we derive their posterior distribution:

$$p\left(C_{yes} \mid x\right) = \frac{p\left(x \mid C_{yes}\right) \cdot p\left(C_{yes}\right)}{p(x)}, \quad p\left(C_{no} \mid x\right) = \frac{p\left(x \mid C_{no}\right) \cdot p\left(C_{no}\right)}{p(x)},$$

where $p(x)$ is the probability density of the newcomer x. In order to identify the class, to which a newcomer x should be assigned, it is reasonable to choose that with the maximal posterior probability. For that, we assume that the prior distribution of the classes is uniform, i.e.

$$p\left(C_{yes}\right) = p\left(C_{no}\right) = \frac{1}{2}.$$

Since the denominator of the posterior distribution is always positive and does not depend on the class, we may equivalently compare the so-called *likelihoods*:

$$L_x\left(C_{yes}\right) = p\left(x \mid C_{yes}\right), \quad L_x\left(C_{no}\right) = p\left(x \mid C_{no}\right).$$

The *maximum likelihood rule* would assign x to the class with the maximum likelihood:

$$y = \begin{cases} +1, & \text{if } L_x\left(C_{yes}\right) \geq L_x\left(C_{no}\right), \\ -1, & \text{else.} \end{cases} \tag{$\mathcal{ML}$}$$

How is it possible to estimate *means* and *covariance matrices* by using the given set of data points? For that, we use the sample mean and the sample covariance matrix:

$$\mu_{yes} = x_{yes}, \quad \Sigma_{yes} = \frac{1}{n_{yes}} \cdot \sum_{i \in C_{yes}} \left(x_i - x_{yes}\right) \cdot \left(x_i - x_{yes}\right)^T,$$

$$\mu_{no} = x_{no}, \quad \Sigma_{no} = \frac{1}{n_{no}} \cdot \sum_{i \in C_{no}} \left(x_i - x_{no}\right) \cdot \left(x_i - x_{no}\right)^T.$$

It turns out that the Fisher's discriminant rule is mimics the maximum likelihood rule. To see this, we assume the *homoscedasticity* of Gauss distributions. The latter means that the covariance matrices corresponding to C_{yes} and C_{no} are identical:

$$\Sigma = \Sigma_{yes} = \Sigma_{no}.$$

As an immediate consequence, the matrices Σ and W coincide:

$$\Sigma = \frac{n_{yes} + n_{no}}{n} \cdot \Sigma = \frac{1}{n} \cdot \left(n_{yes} \cdot \Sigma + n_{no} \cdot \Sigma \right) =$$

$$= \frac{1}{n} \cdot \left(\sum_{i \in C_{yes}} (x_i - x_{yes}) \cdot (x_i - x_{yes})^T + \sum_{i \in C_{no}} (x_i - x_{no}) \cdot (x_i - x_{no})^T \right) = W.$$

Moreover, the condition

$$L_x \left(C_{yes} \right) \geq L_x \left(C_{no} \right)$$

in $(\mathcal{ML})$ for labeling x by $+1$, is equivalent to

$$\left(x - \mu_{yes} \right)^T \cdot \Sigma^{-1} \cdot \left(x - \mu_{yes} \right) \leq \left(x - \mu_{no} \right)^T \cdot \Sigma^{-1} \cdot (x - \mu_{no}).$$

By simplifying the latter, we obtain:

$$\left(\mu_{yes} - \mu_{no} \right)^T \cdot \Sigma^{-1} \cdot \left(x - \frac{\mu_{yes} + \mu_{no}}{2} \right) \geq 0.$$

Substituting here $\mu_{yes} = x_{yes}$, $\mu_{no} = x_{no}$, and $\Sigma = W$, we get:

$$\left(x_{yes} - x_{no} \right)^T \cdot W^{-1} \cdot \left(x - \frac{x_{yes} + x_{no}}{2} \right) \geq 0.$$

This is exactly the Fisher's discriminant rule with $a = W^{-1} \cdot \left(x_{yes} - x_{no} \right)$ derived previously:

$$a^T \cdot x \geq a^T \cdot \left(\frac{x_{yes} + x_{no}}{2} \right) = b.$$

We thus deduced that the Fisher's discriminant can be viewed as a kind of maximum likelihood classifier.

4.2.2 Support-Vector Machine

Let us present the support-vector machine for classification, see e.g. Mathar et al. (2020). Its motivation is mainly rooted in geometrical considerations, involving separating hyperplane and maximum margin.

4.2.2.1 Separating Hyperplane

The idea of support-vector machine is to separate the data points $x_i \in \mathbb{R}^m$ with the corresponding labels $y_i \in \{\pm 1\}$, $i = 1, \ldots, n$, by a hyperplane:

$$H = \left\{ x \in \mathbb{R}^m \mid a^T \cdot x - b = 0 \right\}$$

with $a \in \mathbb{R}^m$ and $b \in \mathbb{R}$ to be yet specified. Geometrically, the hyperplane H divides the data space into two parts:

$$H_\geq = \left\{ x \in \mathbb{R}^m \mid a^T \cdot x - b \geq 0 \right\}, \quad H_< = \left\{ x \in \mathbb{R}^m \mid a^T \cdot x - b < 0 \right\}.$$

The class C_{yes} is on the one side, and the class C_{no} is on the other side of the *separating hyperplane H*, see Fig. 4.3, i.e.

$$x_i \in H_\geq \text{ if and only if } y_i = +1, \quad x_i \in H_< \text{ if and only if } y_i = -1.$$

To classify a newcomer $x \in \mathbb{R}^m$, it is thus sufficient to determine on which side of H it lies:

$$y = \begin{cases} +1, & \text{if } x \in H_\geq, \\ -1, & \text{else.} \end{cases} \qquad (\mathcal{SVM})$$

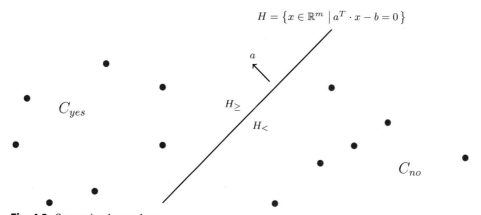

Fig. 4.3 Separating hyperplane

This *support-vector machine rule* (*SVM*) is a linear classifier rule (*LC*):

$$y = \begin{cases} +1, & \text{if } a^T \cdot x \geq b, \\ -1, & \text{else,} \end{cases}$$

where a is the linear classifier and b is the corresponding bound. Now, the crucial question is on how to properly choose the hyperplane H in dependence on the data set $(x_i, y_i) \in \mathbb{R}^m \times \{\pm 1\}, i = 1, \ldots, n$.

4.2.2.2 Maximum Margin

We follow the geometric intuition and select a particular separating hyperplane, which maximizes the distance to the closest data points from each of the classes. For that, let H be an arbitrary separating hyperplane for the classes C_{yes} and C_{no}. Let us carry out a parallel shift of H towards C_{yes} and C_{no} without violating the separation property, see Fig. 4.4. Then, for some $\gamma \geq 0$ and for all $i = 1, \ldots, n$ we have:

$$a^T \cdot x_i - b \geq \gamma \text{ if } y_i = +1, \quad a^T \cdot x_i - b \leq -\gamma \text{ if } y_i = -1.$$

These conditions can be pooled together:

$$y_i \cdot \left(a^T \cdot x_i - b \right) \geq \gamma.$$

Let us consider the region bounded by the two parallel hyperplanes:

$$a^T \cdot x - b = +\gamma, \quad a^T \cdot x - b = -\gamma.$$

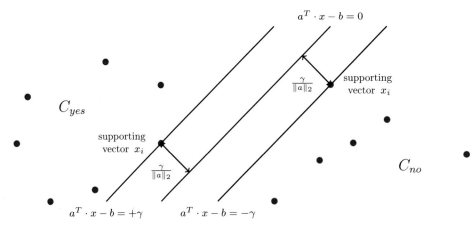

Fig. 4.4 Maximum margin

Its half-width is called *margin*. The margin can be calculated as the distance between these two parallel hyperplanes, see Exercise 4.4:

$$\frac{\gamma}{\|a\|_2},$$

where $\|a\|_2$ denotes the Euclidean norm of the normal vector a. Now, we are ready to state the optimization problem for finding a separating hyperplane with the *maximum margin*:

$$\max_{\gamma \geq 0, a, b} \frac{\gamma}{\|a\|_2} \quad \text{s.t.} \quad y_i \cdot \left(a^T \cdot x_i - b\right) \geq \gamma, \quad i = 1, \dots, n.$$

Since optimal γ's are positive, we may equivalently write:

$$\min_{\gamma > 0, a, b} \left\| \frac{a}{\gamma} \right\|_2 \quad \text{s.t.} \quad y_i \cdot \left(\left(\frac{a}{\gamma}\right)^T \cdot x_i - \frac{b}{\gamma}\right) \geq 1, \quad i = 1, \dots, n.$$

Denoting the scaled variables by a and b again, we eliminate γ:

$$\min_{a, b} \|a\|_2 \quad \text{s.t.} \quad y_i \cdot \left(a^T \cdot x_i - b\right) \geq 1, \quad i = 1, \dots, n.$$

Finally, the objective function can be monotonically transformed:

$$\min_{a, b} \frac{1}{2} \cdot \|a\|_2^2 \quad \text{s.t.} \quad y_i \cdot \left(a^T \cdot x_i - b\right) \geq 1, \quad i = 1, \dots, n. \qquad (\mathcal{P})$$

Note that $(\mathcal{P})$ is a convex optimization problem with a quadratic objective function subject to linear inequality constraints. The difficulty of solving $(\mathcal{P})$ comes from the fact that the number n of constraints is normally huge. In fact, there are as many constraints as data points $i = 1, \dots, n$. Instead of treating the optimization problem $(\mathcal{P})$ directly, let us derive a dual problem which will be easier to tackle.

4.2.2.3 Dual Problem
Let us introduce the Lagrange multipliers

$$\lambda = (\lambda_1, \dots, \lambda_n)^T \in \mathbb{R}^n,$$

and dualize the constraints

$$1 - y_i \cdot \left(a^T \cdot x_i - b\right) \leq 0, \quad i = 1, \dots, n.$$

Then, we equivalently rewrite ($\mathcal{P}$):

$$\min_{a,b} \max_{\lambda \geq 0} \frac{1}{2} \cdot \|a\|_2^2 + \sum_{i=1}^{n} \lambda_i \cdot \left(1 - y_i \cdot \left(a^T \cdot x_i - b\right)\right).$$

If for optimal a, b the i-th constraint is violated, i.e. $1 - y_i \cdot \left(a^T \cdot x_i - b\right) > 0$, then by choosing $\lambda_i \to \infty$, the sum blows up, a contradiction to the minimality of a, b. On the other hand, if the i-th constraint is fulfilled, i.e. $1 - y_i \cdot \left(a^T \cdot x_i - b\right) \leq 0$, then the maximization with respect to the nonnegative Lagrange multipliers provides $\lambda_i = 0$, and the sum vanishes. Now, we apply the *strong duality* of convex optimization, see e.g. Nesterov (2018). In analogy to linear programming, we may exchange minimization and maximization above:

$$\max_{\lambda \geq 0} \min_{a,b} \frac{1}{2} \cdot \|a\|_2^2 + \sum_{i=1}^{n} \lambda_i \cdot \left(1 - y_i \cdot \left(a^T \cdot x_i - b\right)\right).$$

The necessary optimality condition of the inner minimization reads:

$$\nabla_{a,b} \left(\frac{1}{2} \cdot \|a\|_2^2 + \sum_{i=1}^{n} \lambda_i \cdot \left(1 - y_i \cdot \left(a^T \cdot x_i - b\right)\right)\right) = 0,$$

or, equivalently:

$$a = \sum_{i=1}^{n} \lambda_i \cdot y_i \cdot x_i, \quad \sum_{i=1}^{n} \lambda_i \cdot y_i = 0.$$

Substituting into the objective function, we get:

$$\frac{1}{2} \cdot \|a\|_2^2 + \sum_{i=1}^{n} \lambda_i - a^T \cdot \underbrace{\sum_{i=1}^{n} \lambda_i \cdot y_i \cdot x_i}_{=a} + \underbrace{\sum_{i=1}^{n} \lambda_i \cdot y_i \cdot b}_{=0} = \sum_{i=1}^{n} \lambda_i - \frac{1}{2} \cdot \left\|\sum_{i=1}^{n} \lambda_i \cdot y_i \cdot x_i\right\|_2^2.$$

Overall, we obtain the dual problem:

$$\max_{\lambda \geq 0} \sum_{i=1}^{n} \lambda_i - \frac{1}{2} \cdot \sum_{i,j=1}^{n} \lambda_i \cdot \lambda_j \cdot y_i \cdot y_j \cdot x_i^T \cdot x_j \quad \text{s.t.} \quad \sum_{i=1}^{n} \lambda_i \cdot y_i = 0. \qquad (\mathcal{D})$$

This is also a convex optimization problem with a quadratic objective function on the nonnegative orthant subject to just one linear equality constraint. However, the huge number n of variables in ($\mathcal{D}$) becomes an issue. By introducing the dual problem, we shifted the numerical difficulty from dealing with many constraints in $(\mathcal{P})$ to dealing with many variables in $(\mathcal{D})$. Nevertheless, as we shall see in a moment, most of the variables $\lambda_i, i = 1, \ldots, n$, will vanish at the optimum.

4.2.2.4 Supporting Vectors

Let us recover a and b to be optimal for $(\mathcal{P})$ out of the solution λ of $(\mathcal{D})$. For that, we define the index set of positive Lagrange multipliers:

$$S = \{i \in \{1, \ldots, n\} \mid \lambda_i > 0\}.$$

The recovery of a is straightforward:

$$a = \sum_{i \in S} \lambda_i \cdot y_i \cdot x_i.$$

In analogy to linear programming, the complementary slackness provides:

$$\lambda_i \cdot \left(1 - y_i \cdot \left(a^T \cdot x_i - b\right)\right) = 0, \quad i = 1, \ldots, n.$$

In particular, it follows for $i \in S$:

$$y_i \cdot \left(a^T \cdot x_i - b\right) = 1.$$

Multiplying by y_i and using $(y_i)^2 = 1$, we get:

$$b = a^T \cdot x_i - y_i.$$

Moreover, we see that the data points x_i, $i \in S$, fulfill the linear constraints of $(\mathcal{P})$ with equality. Recalling the definition of the maximum margin, we conclude that x_i, $i \in S$, are *supporting vectors*, i.e. they have the smallest distance to the separating hyperplane, see Fig. 4.4:

$$H = \left\{x \in \mathbb{R}^m \;\middle|\; a^T \cdot x - b = 0\right\}.$$

This gives the name of *support-vector machine* to this classification technique, developed in 1960s by Vapnik and Chervonenkis, see e.g. Cortes and Vapnik (1995). We conclude that exclusively the supporting vectors—together with the corresponding labels and Lagrange multipliers—enter into the optimal solution of $(\mathcal{P})$. Since their number $|S|$ is usually much less than n, there is a good hope that most of the variables in $(\mathcal{D})$ vanish at the optimum. This facilitates to efficiently solve the dual problem by the *coordinate gradient descent* method from convex optimization, see e.g. Hsieh et al. (2008).

4.2.2.5 Regularization

Up to now, we always assumed that a separating hyperplane between the classes C_{yes} and C_{no} exists. What can be done if this is not the case? Then, we relax the separability requirement by allowing violations of linear inequality constraints in $(\mathcal{P})$. In order to keep these violations small enough, their sum penalizes the objective function. We consider the following regularized optimization problem:

$$\min_{\xi \geq 0, a, b} \frac{1}{2} \cdot \|a\|_2^2 + \underbrace{c \cdot \sum_{i=1}^{n} \xi_i}_{\text{penalization}} \quad \text{s.t.} \quad y_i \cdot \left(a^T \cdot x_i - b\right) \geq \underbrace{1 - \xi_i}_{\text{violation}} , \quad i = 1, \ldots, n.$$

$$(\mathcal{P}_{reg})$$

Here, $c > 0$ is a constant which expresses the importance of the regularization term. It is not hard to derive the dual problem of $(\mathcal{P}_{reg})$, see Exercise 4.5:

$$\max_{c \cdot e \geq \lambda \geq 0} \sum_{i=1}^{n} \lambda_i - \frac{1}{2} \cdot \sum_{i,j=1}^{n} \lambda_i \cdot \lambda_j \cdot y_i \cdot y_j \cdot x_i^T \cdot x_j \quad \text{s.t.} \quad \sum_{i=1}^{n} \lambda_i \cdot y_i = 0. \qquad (\mathcal{D}_{reg})$$

Note that the dual problems $(\mathcal{D})$ and $(\mathcal{D}_{reg})$ differ just by bounding the Lagrange multipliers from above. Thus, $(\mathcal{D}_{reg})$ can be also efficiently solved by coordinate descent methods. The consequence of regularization is that some, but a few data points may lie on the wrong side of the separating hyperplane, see Fig. 4.5.

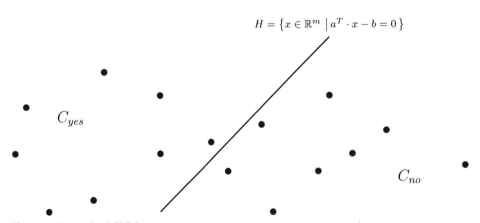

$$H = \left\{ x \in \mathbb{R}^m \mid a^T \cdot x - b = 0 \right\}$$

C_{yes}

C_{no}

Fig. 4.5 Regularized SVM

4.2.2.6 Kernel Trick

Another way to deal with the linear nonseparability of the classes C_{yes} and C_{no} is the kernel trick. Instead of applying the support vector machine to the m-dimensional data directly, they can be first transformed by means of a *feature mapping* $\Phi : \mathbb{R}^m \to \mathbb{R}^r$, where the so-called *intrinsic degree* r denotes the number of new features, see Fig. 4.6. Then, the usual support vector machine is launched on the data set $(\Phi(x_i), y_i) \in \mathbb{R}^r \times \{\pm 1\}$, $i = 1, \ldots, n$. Its dual version is given by $(\mathcal{D})$:

$$\max_{\lambda \geq 0} \sum_{i=1}^{n} \lambda_i - \frac{1}{2} \cdot \sum_{i,j=1}^{n} \lambda_i \cdot \lambda_j \cdot y_i \cdot y_j \cdot \Phi(x_i)^T \cdot \Phi(x_j) \quad \text{s.t.} \quad \sum_{i=1}^{n} \lambda_i \cdot y_i = 0.$$

Now, the crucial question is how to find a feature mapping Φ, so that the transformed data set becomes linearly separable. Exactly here the *kernel trick* helps. The latter suggests to substitute the scalar products by an appropriate kernel, i.e. for $i, j = 1, \ldots, n$ we set:

$$\Phi(x_i)^T \cdot \Phi(x_j) = K(x_i, x_j).$$

The kernilized support vector machine then solves:

$$\max_{\lambda \geq 0} \sum_{i=1}^{n} \lambda_i - \frac{1}{2} \cdot \sum_{i,j=1}^{n} \lambda_i \cdot \lambda_j \cdot y_i \cdot y_j \cdot K(x_i, x_j) \quad \text{s.t.} \quad \sum_{i=1}^{n} \lambda_i \cdot y_i = 0. \quad (\mathcal{D}_{kernel})$$

The intuition behind the kernels is that if $\Phi(x_i)$ and $\Phi(x_j)$ are close, then their scalar product $\Phi(x_i)^T \cdot \Phi(x_j)$ is large. Hence, the kernel $K(x_i, x_j)$ measures *similarity* between x_i and x_j. Moreover, the use of kernels does not make it necessary to explicitly know the feature mapping. This is the case if we guarantee that for a given *kernel* $K : \mathbb{R}^m \times \mathbb{R}^m \to \mathbb{R}$ some feature mapping $\Phi : \mathbb{R}^m \to \mathbb{R}^r$ exists, so that for all $u, v \in \mathbb{R}^m$ it holds:

$$K(u, v) = \Phi(u)^T \cdot \Phi(v).$$

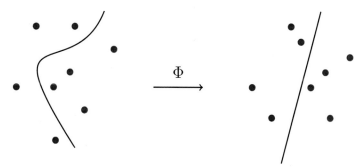

Fig. 4.6 Feature mapping

We call kernels, which can be represented as scalar products of some feature mapping, *valid*. How is it possible to characterize valid kernels? The answer to this question provides Mercer's theorem, see e.g. Kung (2014). In order to state the latter, we define the *kernel* $(n \times n)$-*matrix*:

$$K_X = \left(K \left(x_i, x_j \right) \right)_{i,j},$$

corresponding to a given data sample denoted by

$$X = \left\{ x_i \in \mathbb{R}^m \mid i = 1, \ldots, n \right\}.$$

Mercer's theorem says that a continuous and symmetric kernel K is valid if and only if the kernel matrix K_X is positive semidefinite for all data samples X, i.e. for all $\xi \in \mathbb{R}^n$ it holds:

$$\xi^T \cdot K_X \cdot \xi = \sum_{i,j=1}^{n} \xi_i \cdot \xi_j \cdot K \left(x_i, x_j \right) \geq 0.$$

4.2.2.7 Quadratic Kernel

Exemplarily, we consider the *quadratic kernel*:

$$K(u, v) = \left(u^T \cdot v \right)^2,$$

where $u, v \in \mathbb{R}^m$. Let us show by means of Mercer's theorem that K is a valid. To see this, we compute:

$$\xi^T \cdot K_X \cdot \xi = \sum_{i,j=1}^{n} \xi_i \cdot \xi_j \cdot \left(x_i^T \cdot x_j \right)^2 = \sum_{i,j=1}^{n} \xi_i \cdot \xi_j \cdot \left(\sum_{k=1}^{m} (x_i)_k \cdot (x_j)_k \right)^2$$

$$= \sum_{i,j=1}^{n} \xi_i \cdot \xi_j \cdot \sum_{k,\ell=1}^{m} (x_i)_k \cdot (x_j)_k \cdot (x_i)_\ell \cdot (x_j)_\ell$$

$$= \sum_{k,\ell=1}^{m} \left(\sum_{i=1}^{n} \xi_i \cdot (x_i)_k \cdot (x_i)_\ell \right)^2 \geq 0.$$

Hence, the kernel matrix K_X is positive definite. Due to Mercer's theorem, the quadratic kernel K is thus valid, i.e. it can be represented as follows:

$$K(u, v) = \Phi(u)^T \cdot \Phi(v).$$

A corresponding feature mapping $\Phi : \mathbb{R}^m \to \mathbb{R}^r$ can be given explicitly:

$$\Phi(u) = \left(u_1^2, \ldots, u_m^2, \sqrt{2} \cdot u_1 \cdot u_2, \ldots, \sqrt{2} \cdot u_{m-1} \cdot u_m\right)^T,$$

where the intrinsic degree is

$$r = \binom{m+1}{2} = \frac{m \cdot (m+1)}{2}.$$

We conclude that the use of the quadratic kernel in support vector machine allows to nonlinearly separate the classes C_{yes} and C_{no} by parabolic, hyperbolic, or elliptic surfaces rather than only by hyperplanes. The price to pay is the growing number of features from m up to r by an order of magnitude.

4.3 Case Study: Quality Control

Analytical *quality control* ensures that the results of laboratory analysis are consistent, comparable, accurate and within specified limits of precision. It is aimed to detect, reduce, and correct deficiencies in a laboratory's internal analytical process prior to the release of patient results. In other words, analytical quality control addresses the question on how well the measurement system reproduces the same result over time and under varying operating conditions. In circumstances where more than one laboratory is analyzing samples and feeding data into a large program of work, analytical quality control can also be applied to validate one laboratory against another. In such cases we may refer to *inter-laboratory calibration*. To become concrete, let m laboratories test patients' samples on a virus. The data set consists of their test results $x_i \in \{0, 1\}^m$, and the true diagnosis $y_i \in \{0, 1\}$ of the i-th patient, $i = 1, \ldots, n$. Here, 0 stands for "negative" (not virus-infected), and 1 stands for "positive" (virus-infected). It is statistically estimated that the probability within a population of being virus-infected is $\pi \in (0, 1)$. We intend to construct a so-called *naïve Bayes classifier*, in order to decide whether the incoming test results $x \in \{0, 1\}^m$ of a new patient are more likely to be labeled by either $y = 0$ or $y = 1$.

Task 1 For the j-th laboratory, estimate the probabilities p_j and q_j of the false positive and false negative test results, respectively. By assuming the multivariate Bernoulli event model, determine the conditional probabilities $\mathbb{P}(x \mid y = 0)$ and $\mathbb{P}(x \mid y = 1)$ of the test result $x \in \{0, 1\}^m$ for a not virus-infected and a virus-infected patient, respectively.

Hint 1 The false positive and false negative probabilities are:

$$p_j = \frac{\#\{i \mid (x_i)_j = 1, y_i = 0\}}{\#\{i \mid y_i = 0\}}, \quad q_j = \frac{\#\{i \mid (x_i)_j = 0, y_i = 1\}}{\#\{i \mid y_i = 1\}}.$$

The conditional probabilities are:

$$\mathbb{P}\left(x \mid y = 0\right) = \prod_{j=1}^{m} p_j^{x_j} \cdot \left(1 - p_j\right)^{1-x_j}, \quad \mathbb{P}\left(x \mid y = 1\right) = \prod_{j=1}^{m} \left(1 - q_j\right)^{x_j} \cdot q_j^{1-x_j}.$$

Task 2 By applying *Bayes theorem*, derive the conditional probabilities $\mathbb{P}\left(y = 0 \mid x\right)$ and $\mathbb{P}\left(y = 1 \mid x\right)$ for a patient with the test results x to be not virus-infected or virus-infected, respectively.

Hint 2 Bayes theorem provides:

$$\mathbb{P}\left(y = 0 \mid x\right) = \frac{\mathbb{P}\left(x \mid y = 0\right) \cdot \mathbb{P}\left(y = 0\right)}{\mathbb{P}\left(x\right)}, \quad \mathbb{P}\left(y = 1 \mid x\right) = \frac{\mathbb{P}\left(x \mid y = 1\right) \cdot \mathbb{P}\left(y = 1\right)}{\mathbb{P}\left(x\right)}.$$

Task 3 The *naïve Bayes rule* for a patient with the test results x is as follows:

$$y = \begin{cases} 0, \text{ if } \mathbb{P}\left(y = 0 \mid x\right) \geq \mathbb{P}\left(y = 1 \mid x\right), \\ 1, \text{ else.} \end{cases} \tag{$\mathcal{NB}$}$$

Represent the naïve Bayes rule ($\mathcal{NB}$) as a linear classifier rule ($\mathcal{LC}$).

Hint 3 The condition

$$\mathbb{P}\left(y = 0 \mid x\right) \geq \mathbb{P}\left(y = 1 \mid x\right)$$

of labeling x by $y = 0$, is equivalent to

$$\prod_{j=1}^{m} p_j^{x_j} \cdot \left(1 - p_j\right)^{1-x_j} \cdot \left(1 - \pi\right) \geq \prod_{j=1}^{m} \left(1 - q_j\right)^{x_j} \cdot q_j^{1-x_j} \cdot \pi.$$

After taking the logarithm and rearranging, we get:

$$\sum_{j=1}^{m} x_j \cdot \ln \frac{p_j}{1 - p_j} \cdot \frac{q_j}{1 - q_j} \geq \ln \frac{\pi}{1 - \pi} + \sum_{j=1}^{m} \ln \frac{q_j}{1 - p_j}.$$

This corresponds to the linear classifier rule:

$$y = \begin{cases} 0, \text{ if } a^T \cdot x \geq b, \\ 1, \text{ else,} \end{cases}$$

where the naïve Bayes classifier is

$$a = \left(\ln \frac{p_j}{1 - p_j} \cdot \frac{q_j}{1 - q_j}, j = 1, \ldots, m \right)^T,$$

and the bound is

$$b = \ln \frac{\pi}{1 - \pi} + \sum_{j=1}^{m} \ln \frac{q_j}{1 - p_j}.$$

Task 4 Assume that all laboratories are identically performing, i.e. $p = p_j$ and $q = q_j$ for all $j = 1, \ldots, m$. Moreover, the shares of false positive and false negative results are less than 50 %. How many positive test results j are necessary in order to label a patient as virus-infected according to the naïve Bayes rule?

Hint 4 For labeling a patient by $y = 1$, the number of positive test results should satisfy:

$$j > \frac{\ln \frac{\pi}{1-\pi} + m \cdot \ln \frac{q}{1-p}}{\ln \frac{p}{1-p} \cdot \frac{q}{1-q}}.$$

Task 5 Let us model the beginning of a pandemics where the reliability of tests is low, i.e. false negative results occur extremely often. We assume that 80 % of the population do not have any symptoms. Further, the share of false positive results is 1 %, and of false negative 40 %. Are three positive tests out of ten already enough to hospitalize a patient?

Hint 5 For $\pi = 0.2$, $p = 0.01$, $q = 0.4$, and $m = 10$ we have $j \geq 2.09$. Hence, a patient with just three positive tests is more likely to be virus-infected than not.

4.4 Exercises

Exercise 4.1 (Fisher's Discriminant) The creditworthiness of clients with the following incomes, debts, and salaries in thousand USD is known:

	Client 1	Client 2	Client 3	Client 4	Client 5	Client 6
Income	5	6	7	1	2	3
Debts	10	20	0	30	20	40
Salary	2	3	1	2	4	2
Creditworthiness	Yes	Yes	Yes	No	Yes	No

By using the Fisher's discriminant rule, decide if a newcomer with income of 4, debts of 10, and salary of 1,000 USD should be granted a loan.

Exercise 4.2 (Sample Means) Show the validity of the following formula:

$$\sum_{i \in C_{yes}} (z_i - z_{yes}) \cdot (z_{yes} - \bar{z}) + \sum_{i \in C_{no}} (z_i - z_{no}) \cdot (z_{no} - \bar{z}) = 0,$$

where the sample means are

$$z_{yes} = \frac{1}{n_{yes}} \cdot \sum_{i \in C_{yes}} z_i, \quad z_{no} = \frac{1}{n_{no}} \cdot \sum_{i \in C_{no}} z_i, \quad \bar{z} = \frac{1}{n} \cdot \sum_{i=1}^{n} z_i.$$

Exercise 4.3 (Homogeneous Functions) Let $f, g : \mathbb{R}^m \to \mathbb{R}$ be homogeneous functions of the same degree $\alpha > 0$, i. .e. for all $t > 0$ it holds:

$$f(t \cdot a) = t^\alpha \cdot f(a), \quad g(t \cdot a) = t^\alpha \cdot g(a).$$

If additionally $g(a) > 0$ for all $a \in \mathbb{R}^m$, then

$$\max_a \frac{f(a)}{g(a)} = \max_a \ \{f(a) \mid g(a) = 1\} \, .$$

Apply this result in order to show the equivalence of the optimization problems $(\mathcal{V})$ and $(\mathcal{E})$.

Exercise 4.4 (Hyperplanes) Let two parallel hyperplanes be given:

$$a^T \cdot x - b_1 = 0, \quad a^T \cdot x - b_2 = 0,$$

where $a \in \mathbb{R}^m$ and $b_1, b_2 \in \mathbb{R}$. Show that the distance between them equals to

$$\frac{|b_1 - b_2|}{\|a\|_2} \, .$$

Apply this result to derive the formula for the margin.

Exercise 4.5 (Regularized SVM) Show that $(\mathcal{D}_{reg})$ is dual for the regularized optimization problem $(\mathcal{P}_{reg})$. Recover the solution of $(\mathcal{P}_{reg})$ out of the solution of $(\mathcal{D}_{reg})$.

Exercise 4.6 (Kernel Rules) Let $K : \mathbb{R}^m \times \mathbb{R}^m \to \mathbb{R}$ and $L : \mathbb{R}^m \times \mathbb{R}^m \to \mathbb{R}$ be valid kernels with intrinsic degrees r_K and r_L, respectively. Show that the following kernels are also valid:

$$S(u, v) = K(u, v) + L(u, v), \quad P(u, v) = K(u, v) \cdot L(u, v).$$

For their intrinsic degrees show:

$$r_S = r_K + r_L, \quad r_P = r_K \cdot r_L.$$

Exercise 4.7 (Polynomial Kernel) Show that the *polynomial kernel* is valid:

$$K(u, v) = \left(u^T \cdot v\right)^d,$$

where $u, v \in \mathbb{R}^m$ and $d \in \mathbb{N}$. Construct the corresponding feature mapping and compute the intrinsic degree.

Clustering

5

Clustering aims to group a set of objects in such a way that objects within one and the same cluster are more similar to each other than to those in other clusters. Depending on the objects' features, the clustering of DNA sequences of genes, members within a social network, texts written in natural languages, time series of stock prices, medical images from computer tomography, or consumer products on e-commerce platforms, may become relevant. Clustering by itself is not a specific algorithm, but rather a task to be solved. It can be achieved by various algorithms that differ significantly in their understanding of what constitutes a cluster and how to efficiently identify them. In this chapter, we shall present the celebrated *k-means clustering* based on a general *dissimilarity measure* between the objects. In the first step, the algorithm assigns each object to the cluster with the least dissimilar center. In the second step, the centers are recalculated by minimizing the dissimilarity within the clusters. The *k*-means algorithm is specified for the Euclidean setup, where centers turn out to be clusters' sample means. Additionally, we discuss the modifications of *k*-means with respect to other dissimilarity measures. They include *Levenshtein distance, Manhattan norm, cosine similarity, Pearson correlation* and *Jaccard coefficient*. Finally, the technique of *spectral clustering* is used for community detection. It is based on the diffusion of information through a social network and the spectral analysis of the corresponding matrix of transition probabilities.

5.1 Motivation: DNA Sequencing

In recent years, vast collections of *gene expression data* has become available. In particular, the sequencing of DNA has attracted a lot of attention. In 1998, the Icelandic Parliament Althing decided to allow a biopharmaceutical company "deCODE genetics" to collect and store all health data of the population nationwide. The legal basis has been the

© Springer-Verlag GmbH Germany, part of Springer Nature 2021
V. Shikhman, D. Müller, *Mathematical Foundations of Big Data Analytics*,
https://doi.org/10.1007/978-3-662-62521-7_5

then approved Act on Biobanks. Already in 2003, deCODE launched an online version of the DNA sequencing database, called Íslendingabók, or the Book of Icelanders. Anyone with an Icelandic social security number could research their family tree and see their nearest family connection to anyone else in the country. By 2020, it had over 200,000 registered users and more than 900,000 linked entries of DNA sequences, comprising the majority of Icelanders who have ever lived. Since the deCODE gene expression data sets have remained among the largest and best powered collections all over the world, the special emphasis has been placed on the analysis of genealogical relationships. In particular, searching for meaningful information patterns and dependencies in gene expression data, being a difficult, but indispensable task, provides a basis for hypothesis testing. A solution to this problem is to relate DNA sequences to each other, rather than to revisit every new DNA sequence independently. The identification of homogeneous groups with similar DNA patterns is usually done by *clustering*. The clustering of DNA sequences has been proven to be useful for revealing natural structures inherent in gene expression data, which includes gene functions, cellular processes and types. The other benefit is the better understanding of gene homology, which is very important in vaccine design. The clustering of gene expression data crucially depends on a dissimilarity measure for the corresponding DNA's. To put it simply, a DNA strand can be represented as a string, i.e. a finite sequence of characters from the alphabet:

$$\Sigma = \{\text{Adenin (A), Guanin (G), Thymin (T), Cytosin (C)}\}.$$

The set of strings of length s is denoted by Σ^s, and the set of all finite strings (regardless of their length) is indicated by the Kleene star operator:

$$\Sigma^* = \bigcup_{s \in \mathbb{N} \cup \{0\}} \Sigma^s.$$

For two DNA sequences $x, y \in \Sigma^*$ we have to define the corresponding dissimilarity $d(x, y)$. For this purpose it is reasonable to take the Levenshtein distance, which measures the editing difference between the two sequences:

$$d(x, y) = \text{lev}(x, y).$$

The *Levenshtein distance* $\text{lev}(x, y)$ is the minimum number of single-character edits (insertions, deletions or substitutions) required to change the string x into the string y. It is named after Levenshtein (1966). In order to understand this definition let us compute the Levenshtein distance between

$$x = \text{ACCGAT}, \quad y = \text{AGCAT}.$$

One possibility to obtain y from x by editing is to substitute the second character C by G, the fourth character G by A, the fifth character A by T, and to delete the last character T:

$$
\begin{array}{cccccc}
\text{A} & \text{C} & \text{C} & \text{G} & \text{A} & \text{T} \\
& \uparrow & & \uparrow & \uparrow & \uparrow \\
& \text{G} & & \text{A} & \text{T} & \times
\end{array}
$$

Altogether, four edits were needed, but we can do even better. In fact, let us just substitute the second character C by G, and delete the fourth character G:

$$
\begin{array}{cccccc}
\text{A} & \text{C} & \text{C} & \text{G} & \text{A} & \text{T} \\
& \uparrow & & \uparrow & & \\
& \text{G} & & \times & &
\end{array}
$$

Hence, for the Levenshtein distance it holds: $\text{lev}(x, y) = 2$. Now, we turn our attention to the general question on how to efficiently cluster objects by means of a given dissimilarity measure?

5.2 Results

5.2.1 k-Means

We start by describing and analyzing the basic k-means algorithm for clustering.

5.2.1.1 Total Dissimilarity

Let n objects $x_1, \ldots, x_n$ from a data set be given. They need to be grouped into k homogeneous clusters $C_1, \ldots, C_k$, whose disjoint union forms a partition, i.e.

$$
\bigcup_{\ell=1}^{k} C_\ell = \{1, \ldots, n\}.
$$

Homogeneity of a cluster means that its members are close to each other. At the same time, the members of different clusters should be significantly discriminable. To capture this idea, we assume that *dissimilarity $d(x, y)$* between two arbitrary objects x, y can be efficiently measured. The latter allows to perform clustering. For that, we associate with clusters $C_1, \ldots, C_k$ their centers $z_1, \ldots, z_k$. The center z_ℓ is meant to be a representative of the cluster C_ℓ, so that the dissimilarity within the cluster can be computed as

$$d\left(C_\ell, z_\ell\right) = \sum_{i \in C_\ell} d\left(x_i, z_\ell\right).$$

The total dissimilarity is then given by

$$d\left(C, z\right) = \sum_{\ell=1}^{k} d\left(C_\ell, z_\ell\right),$$

where the clusters and the corresponding centers are denoted by

$$C = (C_1, \ldots, C_k), \quad z = (z_1, \ldots, z_k).$$

Our goal is to find a partition C of clusters and their centers z with the minimal total dissimilarity:

$$\min_{C,z} \ d(C, z). \tag{$\mathcal{D}$}$$

5.2.1.2 Naïve k-Means

The optimization problem $(\mathcal{D})$ is rather hard to tackle due to its combinatorial nature. Nevertheless, already a simple iteration scheme, which alternates the updates of clusters and centers, turns out to be helpful. The corresponding algorithm has been proposed by Lloyd (1982) from Bell Labs as a technique for pulse-code modulation, though the idea goes back already to Steinhaus (1957). Let us present the *k-means clustering*:

(1) Next *clusters* consist of objects with the least dissimilarity to the previous centers, i.e. for all $\ell = 1, \ldots, k$ it holds:

$$C_\ell(t + 1) = \left\{i \in \{1, \ldots, n\} \ \big| \ d\left(x_i, z_\ell(t)\right) \le d\left(x_i, z_{\ell'}(t)\right) \text{ for all } \ell' = 1, \ldots, k\right\}.$$

(2) Next *centers* are chosen to minimize the dissimilarity within the newly defined clusters, i.e. for all $\ell = 1, \ldots, k$ it holds:

$$z_\ell(t + 1) \in \arg\min_{z} \ d\left(C_\ell(t + 1), z\right).$$

The k-means clustering consists of an alternating cluster-center update from the t-th to the $(t + 1)$-th iterate:

$$(C(t), z(t)) \overset{(1)}{\longmapsto} (C(t + 1), z(t)) \overset{(2)}{\longmapsto} (C(t + 1), z(t + 1)).$$

Let us show that the total dissimilarity does not increase:

$$d(C(t+1), z(t+1)) = \sum_{\ell=1}^{k} d\left(C_\ell(t+1), z_\ell(t+1)\right) \overset{(2)}{\leq} \sum_{\ell=1}^{k} d\left(C_\ell(t+1), z_\ell(t)\right)$$

$$= \sum_{\ell=1}^{k} \sum_{i \in C_\ell(t+1)} d\left(x_i, z_\ell(t)\right) \overset{(1)}{\leq} \sum_{\ell'=1}^{k} \sum_{i \in C_{\ell'}(t)} d\left(x_i, z_{\ell'}(t)\right)$$

$$= \sum_{\ell'=1}^{k} d\left(C_{\ell'}(t), z_{\ell'}(t)\right) = d(C(t), z(t)).$$

Due to the finite number of partitions, the sequence $(d(C(t), z(t)))_{t \in \mathbb{N}}$ of total dissimilarities has a finite number of nonincreasing values. Hence, k-means stops after a finite number of steps providing a *local* minimizer of the optimization problem $(\mathcal{D})$. The convergence of k-means towards *global* minimizers crucially depends on the initialization of the centers, and cannot be expected to hold in general. We see by example in Fig. 5.1 that, unless the centers are initialized properly, the global convergence of the k-means algorithm may fail, at least in the Euclidean setup. To overcome this obstacle, it is convenient to initialize k-means by a different set of centers repeatedly. Another initialization strategy called *k-means++* has been suggested by Arthur and Vassilvitskii (2007). There, the centers are picked at random—with probability proportional to the squared distance from the centers already chosen.

5.2.1.3 Euclidean Setup

We consider the k-means clustering in the Euclidean setup. For that, we assume that the data points $x_1, \ldots, x_n \in \mathbb{R}^m$ describe objects' m features, and the dissimilarity measure between two objects $x, y \in \mathbb{R}^m$ is given by the *Euclidean distance*:

$$d(x, y) = \|x - y\|_2^2.$$

Let us study the structure of clusters and centers given by (1) and (2), respectively. For simplicity we omit the iteration number $t + 1$. First, we make the update (2) of centers explicit. For that, we need to find the center z_ℓ, which minimizes the dissimilarity within the cluster C_ℓ:

$$\min_z \sum_{i \in C_\ell} \|x_i - z\|_2^2.$$

The necessary optimality condition reads:

$$\nabla_z \left(\sum_{i \in C_\ell} \|x_i - z\|_2^2 \right) = 0,$$

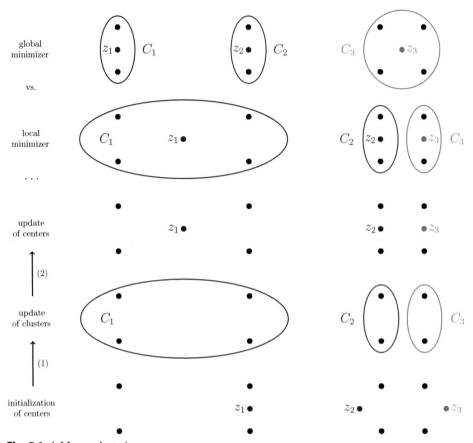

Fig. 5.1 k-Means clustering

or, equivalently:

$$2 \cdot \sum_{i \in C_\ell} (z - x_i) = 0.$$

This provides the well-known formula of the *sample mean*:

$$z_\ell = \frac{1}{|C_\ell|} \sum_{i \in C_\ell} x_i,$$

where $|C_\ell|$ denotes the number of members within C_ℓ. We conclude that the centers from (2) correspond to the sample means of the data points within the clusters. This is where the term k-means originates from. Further, note that the centers induce a so-called *Voronoi diagram*: a decomposition of $\mathbb{R}^m$ into k convex cells, each of them containing the region of

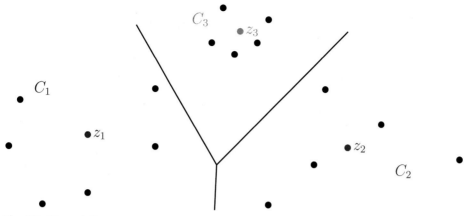

Fig. 5.2 Voronoi diagram

space whose nearest center is z_ℓ, $\ell = 1, \ldots, k$. Since the clusters from (1) have the form:

$$C_\ell = \left\{ i \in \{1, \ldots, n\} \mid \|x_i - z_\ell\|_2 \leq \|x_i - z_{\ell'}\|_2 \text{ for all } \ell' = 1, \ldots, k \right\},$$

the data points x_i, $i \in C_\ell$, lie within one and the same Voronoi cell, see Fig. 5.2.

5.2.2 Spectral Clustering

As an application of k-means, let us consider the spectral clustering for community detection, see e.g. Mathar et al. (2020).

5.2.2.1 Community Detection

Community detection is a key to understanding the structure of complex social networks, and ultimately extracting useful information from them. The motives behind community detection are diverse. It can help a brand to identify different groups of opinion toward its products. Targeting certain groups of people, who have similar interests and are geographically near to each other, may lead to tailored solutions and improve the performance of services provided. Identifying similar customers in the network of purchase relationships enables online retailers to set up efficient recommendations, that better guide customers through the list of items and enhance the business opportunities of retailers. Let us mathematically model the problem of community detection. For that, we assume that there are n persons communicating within a *social network*, see Fig. 5.3. For every two of them i and j, let their *linkage* $w_{ij} \geq 0$ be known. The linkage w_{ij} weights the intensity of the contacts between the persons i and j. This can be e.g. an average duration of mutual daily

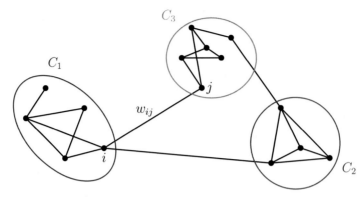

Fig. 5.3 Community detection

calls, so that for all $i, j = 1, \ldots, n$ it holds:

$$w_{ij} = w_{ji}.$$

For brevity, we define the symmetric linkage $(n \times n)$-matrix with nonnegative entries:

$$W = \left(w_{ij}\right).$$

In order to cluster persons into k groups, an appropriate dissimilarity measure is needed. Often, it is not adequate to use the linkage directly. In fact, it may happen that people talk a lot to each other, but arrive at quite different opinions due to other persons in contact. The idea of spectral clustering is to rather look at the *diffusion of information* within a social network. Loosely speaking, two persons are similar if they cause more or less the same information footstep on the social network after a while. In other words, the dissimilarity of persons is measured by their ability to manipulate other people's opinions in different ways. Let us cast this idea in mathematical terms.

5.2.2.2 Diffusion of Information
First, we introduce the probabilities of information diffusion from person j to i:

$$p_{ij} = \frac{w_{ij}}{\sum_{i=1}^{n} w_{ij}}.$$

For the diffusion $(n \times n)$-matrix $P = \left(p_{ij}\right)$ it holds:

$$P = W \cdot D^{-1},$$

where the diagonal matrix of degrees is given by

$$D = \text{diag}\left(\sum_{i=1}^{n} w_{ij}, j = 1, \ldots, n\right).$$

Note that P is a *stochastic matrix* by construction, i.e. its entries are nonnegative and its columns sum up to one:

$$P \geq 0, \quad e^T \cdot P = e^T,$$

where $e = (1, \ldots, 1)^T$ stands for the n-dimensional vector of ones. Now, we are ready to describe the process of information diffusion. For that, let the j-th person start with a gossip:

$$x_j(0) = e_j,$$

where e_j is the j-th coordinate vector of $\mathbb{R}^n$, i.e. the j-th entry of e_j equals to one, and zero otherwise. This means that all people within the social network, but the j-th person, have yet no clue about the news. After passing to j-th person's friends, the information spreads over the whole social network from one contact to another. This diffusion of information is thus given by the update

$$x_j(t) = P \cdot x_j(t - 1),$$

where $x_j(t - 1) \in \Delta$ can be viewed as an information footstep caused by j-th person on the social network at time $t - 1$. Recall that we use here the simplex:

$$\Delta = \left\{x \in \mathbb{R}^n \mid x \geq 0 \text{ and } e^T \cdot x = 1\right\}.$$

The proposed dynamics is well-defined, since at time zero the j-th person starts with the information footstep $x_j(0) \in \Delta$. Then, $x(t)$ becomes also an information footstep on the social network, i.e. $x_j(t) \in \Delta$. This can be shown by induction. Since the matrix P is stochastic, we have namely:

$$x_j(t) = \underbrace{P}_{\geq 0} \cdot \underbrace{x_j(t - 1)}_{\geq 0} \geq 0,$$

and

$$e^T \cdot x_j(t) = \underbrace{e^T \cdot P}_{=e^T} \cdot x_j(t - 1) = e^T \cdot x_j(t - 1) = 1.$$

Explicitly, the diffusion of information can be written as

$$x_j(t) = P \cdot \underbrace{x_j(t-1)}_{=P \cdot x_j(t-2)} = P^2 \cdot x_j(t-2) = \ldots = P^t \cdot x_j(0) = P^t \cdot e_j.$$

In order to compute the t-th power of the diffusion matrix P, we derive its spectral decomposition.

5.2.2.3 Spectral Decomposition

We start with the diagonalization of the auxiliary matrix:

$$S = D^{-1/2} \cdot W \cdot D^{-1/2},$$

where

$$D^{-1/2} = \text{diag} \left(\frac{1}{\sqrt{\sum_{i=1}^{n} w_{ij}}}, j = 1, \ldots, n \right).$$

Note that the matrix S is symmetric, since for its transpose holds:

$$S^T = \left(D^{-1/2} \cdot W \cdot D^{-1/2} \right)^T = \left(D^{-1/2} \right)^T \cdot W^T \cdot \left(D^{-1/2} \right)^T = D^{-1/2} \cdot W \cdot D^{-1/2} = S.$$

We consider the eigenvectors $v_1, \ldots, v_n \in \mathbb{R}^n$ of the symmetric matrix S, which correspond to its eigenvalues $\lambda_1, \ldots, \lambda_n \in \mathbb{R}$. By definition, it holds for all $r = 1, \ldots, n$:

$$S \cdot v_r = \lambda_r \cdot v_r.$$

Equivalently, we have in matrix form:

$$S \cdot V = V \cdot \Lambda,$$

where the columns of the matrix $V = (v_1, \ldots, v_n)$ are formed by the eigenvectors of S, and the diagonal matrix $\Lambda = \text{diag}(\lambda_1, \ldots, \lambda_n)$ has the eigenvalues of S on its diagonal. For simplicity, we assume that all eigenvalues of S are pairwise different, and sorted in decreasing order:

$$|\lambda_1| > \ldots > |\lambda_n|.$$

Then, we have:

$$\lambda_s \cdot v_r^T \cdot v_s = v_r^T \cdot \underbrace{\lambda_s \cdot v_s}_{=S \cdot v_s} = v_r^T \cdot S \cdot v_s = \left(S^T \cdot v_r\right)^T \cdot v_s = \underbrace{(S \cdot v_r)^T}_{=\lambda_r \cdot v_r^T} \cdot v_s = \lambda_r \cdot v_r^T \cdot v_s.$$

Since $\lambda_r \neq \lambda_s$ by assumption, we deduce from here that $v_r^T \cdot v_s = 0$ for all $r \neq s$. Without loss of generality, we may additionally assume that the eigenvectors are normalized with respect to the Euclidean norm, i.e. $\|v_r\|_2 = 1$ for all $r = 1, \ldots, n$. Altogether, the eigenvectors of S are pairwise orthogonal:

$$v_r^T \cdot v_s = \begin{cases} 1, & \text{if } r = s, \\ 0, & \text{if } r \neq s. \end{cases}$$

Equivalently, V is an *orthogonal matrix*:

$$V^T \cdot V = V \cdot V^T = I,$$

where I denotes the identity matrix. Overall, we diagonalize:

$$S = S \cdot \underbrace{V \cdot V^T}_{=I} = \underbrace{S \cdot V}_{=V \cdot \Lambda} \cdot V^T = V \cdot \Lambda \cdot V^T.$$

Now, we are ready to represent the diffusion matrix P in spectral form:

$$P = W \cdot D^{-1} = \underbrace{D^{1/2} \cdot D^{-1/2}}_{=I} \cdot W \cdot \underbrace{D^{-1/2} \cdot D^{-1/2}}_{=D^{-1}} = D^{1/2} \cdot \underbrace{D^{-1/2} \cdot W \cdot D^{-1/2}}_{=S} \cdot D^{-1/2}$$

$$= D^{1/2} \cdot \underbrace{S}_{=V \cdot \Lambda \cdot V^T} \cdot D^{-1/2} = \underbrace{D^{1/2} \cdot V}_{=\Phi} \cdot \Lambda \cdot \underbrace{V^T \cdot D^{-1/2}}_{=\Psi^T} = \Phi \cdot \Lambda \cdot \Psi^T.$$

The matrices $\Psi = (\psi_1, \ldots, \psi_n)$ and $\Phi = (\phi_1, \ldots, \phi_n)$ turn out to be *bi-orthonormal*, i.e. their columns are orthogonal to each other:

$$\Psi^T \cdot \Phi = \underbrace{V^T \cdot D^{-1/2}}_{=\Psi^T} \cdot \underbrace{D^{1/2} \cdot V}_{=\Phi} = V^T \cdot \underbrace{D^{-1/2} \cdot D^{1/2}}_{=I} \cdot V = V^T \cdot V = I,$$

or, equivalently:

$$\psi_r^T \cdot \phi_s = \begin{cases} 1, & \text{if } r = s, \\ 0, & \text{if } r \neq s. \end{cases}$$

5.2.2.4 Diffusion Map

By using the spectral decomposition of P, we obtain a simple representation of its t-th power:

$$P^t = \underbrace{\Phi \cdot \Lambda \cdot \Psi^T}_{=P} \cdot \underbrace{\Phi \cdot \Lambda \cdot \Psi^T}_{=P} \cdot \ldots \cdot \underbrace{\Phi \cdot \Lambda \cdot \Psi^T}_{=P} \cdot \underbrace{\Phi \cdot \Lambda \cdot \Psi^T}_{=P}$$

$$= \Phi \cdot \Lambda \cdot \underbrace{\Psi^T \cdot \Phi}_{=I} \cdot \Lambda \cdot \Psi^T \ldots \cdot \Phi \cdot \Lambda \cdot \underbrace{\Psi^T \cdot \Phi}_{=I} \cdot \Lambda \cdot \Psi^T = \Phi \cdot \Lambda^t \cdot \Psi^T.$$

We proceed with the information diffusion caused by the j-th person:

$$x_j(t) = P^t \cdot e_j = \Phi \cdot \Lambda^t \cdot \Psi^T \cdot e_j = \Phi \cdot \Lambda^t \cdot \begin{pmatrix} (\psi_1)_j \\ \vdots \\ (\psi_n)_j \end{pmatrix} = \Phi \cdot \begin{pmatrix} \lambda_1^t \cdot (\psi_1)_j \\ \vdots \\ \lambda_n^t \cdot (\psi_n)_j \end{pmatrix}.$$

By defining the *diffusion map* at time t as

$$F_j(t) = \begin{pmatrix} \lambda_1^t \cdot (\psi_1)_j \\ \vdots \\ \lambda_n^t \cdot (\psi_n)_j \end{pmatrix},$$

we obtain:

$$x_j(t) = \Phi \cdot F_j(t).$$

Now, let us measure the difference between the information footsteps $x_i(t)$ and $x_j(t)$ on the social network caused by the persons i and j, respectively. For that, we use the weighted Euclidean distance:

$$\left\| D^{-1/2} \cdot \left(x_i(t) - x_j(t) \right) \right\|_2^2 = \left\| D^{-1/2} \cdot \Phi \cdot \left(F_i(t) - F_j(t) \right) \right\|_2^2$$

$$= \left\| D^{-1/2} \cdot D^{1/2} \cdot V \cdot \left(F_i(t) - F_j(t) \right) \right\|_2^2$$

$$= \left\| V \cdot \left(F_i(t) - F_j(t) \right) \right\|_2^2$$

$$= \left(V \cdot \left(F_i(t) - F_j(t) \right) \right)^T \cdot \left(V \cdot \left(F_i(t) - F_j(t) \right) \right)$$

$$= \left(F_i(t) - F_j(t) \right)^T \cdot V^T \cdot V \cdot \left(F_i(t) - F_j(t) \right)^T$$

$$= \left(F_i(t) - F_j(t)\right)^T \cdot \left(F_i(t) - F_j(t)\right)$$
$$= \left\| F_i(t) - F_j(t) \right\|_2^2 .$$

We conclude that the dissimilarity between two persons within a social network can be captured by the Euclidean distance of the corresponding diffusion maps. The later is called the *diffusion distance*.

5.2.2.5 Dimensionality Reduction

Since the diffusion distance is an appropriate dissimilarity measure within a social network, we turn our attention to the study of diffusion maps. First, we claim that $\lambda_1, \ldots, \lambda_n$ are eigenvalues of the diffusion matrix P, and the corresponding eigenvectors are the columns $\phi_1, \ldots, \phi_n$ of the matrix Φ. In fact, we have for all $s = 1, \ldots, n$:

$$P \cdot \phi_s = \Phi \cdot \Lambda \cdot \underbrace{\Psi^T \cdot \phi_s}_{=e_s} = \Phi \cdot \Lambda \cdot e_s = \lambda_s \cdot \Phi \cdot e_s = \lambda_s \cdot \phi_s.$$

Since all eigenvalues of a stochastic matrix are bounded by one, see Exercise 5.3, and there exists at least one of them being equal to one, we get:

$$\lambda_1 = 1.$$

Note that the eigenvector ϕ_1 is a ranking, see Chap. 1. Second, we show that the columns $\psi_1, \ldots, \psi_n$ of the matrix Ψ are eigenvectors of P^T. In fact, we have for all $r = 1, \ldots, n$:

$$P^T \cdot \psi_r = \left(\Phi \cdot \Lambda \cdot \Psi^T\right)^T \cdot \psi_r = \Psi \cdot \Lambda \cdot \underbrace{\Phi^T \cdot \psi_r}_{=e_r} = \Psi \cdot \Lambda \cdot e_r = \lambda_r \cdot \Psi \cdot e_r = \lambda_r \cdot \psi_r.$$

Since transposing $e^T \cdot P = e^T$ gives $P^T \cdot e = e$, and $\lambda_1 = 1$, we get:

$$\psi_1 = e.$$

Altogether, the first components of the diffusion maps coincide for all $j = 1, \ldots, n$:

$$\left(F_j\right)_1 (t) = \lambda_1^t \cdot (\psi_1)_j = 1.$$

Their last, say $n - k$, components tend to zero if time progresses. This is due to the fact that all eigenvalues of P, but the first one, are strictly bounded by one:

$$|\lambda_\ell| < 1 \quad \text{for all } \ell = k + 1, \ldots, n.$$

Hence, for all $\ell = k + 1, \ldots, n$ and $j = 1, \ldots, n$ we have:

$$\left(F_j\right)_\ell (t) = \lambda_\ell^t \cdot (\psi_\ell)_j \to 0 \quad \text{for } t \to \infty.$$

This motivates the introduction of the *k-truncated diffusion map*:

$$k\text{-}F_j(t) = \begin{pmatrix} \lambda_2^t \cdot (\psi_2)_j \\ \vdots \\ \lambda_k^t \cdot (\psi_k)_j \end{pmatrix} \in \mathbb{R}^{k-1}.$$

Here, we truncated the first and $n - k$ last components of the diffusion map $F_j(t)$, since the effect of their respective comparison within the diffusion distance is either nonexistent or neglectable. The *dimensionality reduction* allows to associate k-truncated diffusion maps with $k - 1$ features to an every person. The dissimilarity measure between two persons thus reduces to the k-truncated diffusion distance. Finally, let us present the procedure of *spectral clustering*:

(i) Choose t sufficiently large and compute k-truncated diffusion maps:

$$k\text{-}F_1(t), \ldots, k\text{-}F_n(t) \in \mathbb{R}^{k-1}.$$

(ii) Apply k-means clustering for these $(k - 1)$-dimensional feature vectors in the Euclidean setup by using the k-truncated diffusion distance as a dissimilarity measure:

$$d\left(k\text{-}F_i(t), k\text{-}F_j(t)\right) = \left\| k\text{-}F_i(t) - k\text{-}F_j(t) \right\|_2^2.$$

5.3   Case Study: Topic Extraction

Topic extraction is a natural language processing technique that allows to automatically extract meaning from texts by identifying recurrent themes. One possibility to perform topic extraction is by *document clustering*. For that, an appropriate representation of documents is needed. Standardly, it is derived from the *bag-of-words model* first mentioned by Zellig (1954). In this model, a document is represented as the set of its terms, disregarding grammar and even term order, but keeping their frequencies within a document and importance within a document collection. Let us cluster the following travel reviews D1–D6 from a holiday portal:

D1: I spent one week at **Lake** Garda, which is beautifully surrounded by **hills**. There were a lot of sport and wellness offers I tried, but most of the time I relaxed at the **beach** and enjoyed the sun.

D2: We had astonishing **beach** holidays at the **sea**. Weather and hotel have been perfect.

D3: The only things I want to do on holidays is relaxing at the water, while catching some sun. Our balcony provided a nice view of the **lake** with some **mountains** behind.

D4: We had a short hiking trip to the **hills** and **mountains** of Austria.

D5: We made a bike tour through the **hills**, because we like the quiet environment. Fortunately, we had also time for a day in the beautiful spa area.

D6: I hate **beaches**, but love **mountains** and snow. The slopes were perfectly prepared for skiing. We had a great day!

The relevant terms are marked in bold:

$$T1 = \text{"lake"}, \quad T2 = \text{"sea"}, \quad T3 = \text{"beach"}, \quad T4 = \text{"hills"}, \quad T5 = \text{"mountains"}.$$

Task 1 The binary *term frequency* accounts for a term T to occur in a document D:

$$tf(T, D) = \begin{cases} 1, \text{ if } T \in D, \\ 0, \text{ else.} \end{cases}$$

Compute the binary term-frequencies for terms T1–T4 and documents D1–D6.

Hint 1 The term-frequencies are stored in the following table:

tf	D1	D2	D3	D4	D5	D6
T1	1	0	1	0	0	0
T2	0	1	0	0	0	0
T3	1	1	0	0	0	1
T4	1	0	0	1	1	0
T5	0	0	1	1	0	1

Task 2 The *inverse document frequency* is a measure of how much information a term provides, i.e. if it's common or rare across all documents. It is the logarithmically scaled inverse fraction of the documents that contain the term T, obtained by dividing the total number of documents n by the number of documents containing it:

$$idf(T) = \log \frac{n}{\#\{D \mid T \in D\}}.$$

Compute the inverse document frequencies for terms T1–T4.

Hint 2 The inverse document frequencies are stored in the following table:

idf	T1	T2	T3	T4	T5
	$\log 3$	$\log 6$	$\log 2$	$\log 2$	$\log 2$

Task 3 The *term frequency-inverse document frequency* reflects how important a term T is to a document D in a collection:

$$tf\text{-}idf(\text{T}, \text{D}) = tf(\text{T}, \text{D}) \cdot idf(\text{T}).$$

The $tf\text{-}idf$ value increases proportionally to the number of times a term appears in the document and is offset by the number of documents that contain the term. This helps to adjust for the fact that some terms appear more frequently in general. Compute the term frequency-inverse document frequency for terms T1–T4 and documents D1–D6. Represent documents D1–D6 by their $tf\text{-}idf$ values $x_1, \ldots, x_6 \in \mathbb{R}^5$ over the terms.

Hint 3 The term frequency-inverse document frequency values are stored in the following table, whose columns are the $tf\text{-}idf$ representations $x_1, \ldots, x_6$ of documents D1–D6:

$tf\text{-}idf$	D1	D2	D3	D4	D5	D6
T1	$\log 3$	0	$\log 3$	0	0	0
T2	0	$\log 6$	0	0	0	0
T3	$\log 2$	$\log 2$	0	0	0	$\log 2$
T4	$\log 2$	0	0	$\log 2$	$\log 2$	0
T5	0	0	$\log 2$	$\log 2$	0	$\log 2$

Task 4 The *cosine similarity* between $tf\text{-}idf$ representations x, y of documents is defined as

$$\text{Cosine}(x, y) = \frac{x^T \cdot y}{\|x\|_2 \cdot \|y\|_2}.$$

Let the dissimilarity measure for documents be taken as

$$d(x, y) = \frac{1 - \text{Cosine}(x, y)}{2}.$$

Derive a formula for the update (2) of centers in k-means based on the cosine similarity.

Hint 4 In (2), we need to find the center z_ℓ, which minimizes the dissimilarity within the cluster C_ℓ:

$$\min_z \sum_{i \in C_\ell} \frac{1 - \text{Cosine}(x_i, z)}{2}.$$

Due to Exercise 4.3, the latter is equivalent to

$$\max_{\|z\|_2 = 1} \left(\sum_{i \in C_\ell} \frac{x_i}{\|x_i\|_2} \right)^T \cdot z.$$

By means of the Lagrange multiplier rule, see e.g. Jongen et al. (2004), we obtain:

$$z_\ell = \sum_{i \in C_\ell} \frac{x_i}{\|x_i\|_2}.$$

Task 5 Apply the k-means algorithm in order to cluster documents D1–D6 into two groups. Use the cosine-based dissimilarity of their $tf\text{-}idf$ representations $x_1, \ldots, x_6 \in \mathbb{R}^5$. Try to initialize the k-means randomly. Can you extract topics from this clustering?

Hint 5 The application of k-means provides:

$$C_1 = \{1, 3, 4, 5\}, \quad C_2 = \{2, 6\}.$$

The relevant topics can be "water" for the documents D1–D3, D5 from the cluster C_1, and "hiking" for the documents D4, D6 from the cluster C_2. Note that our choice of terms neglects the negative attitude towards the beaches in the document D6. It is worth to mention that more data samples would lead to a better interpretation.

5.4 Exercises

Exercise 5.1 (k-Means Clustering) The following data points are given:

$$x_1 = (1, 0)^T, \quad x_2 = (2, 0)^T, \quad x_3 = (3, 0)^T, \quad x_4 = (4, 0)^T, \quad x_5 = (5, 0)^T, \quad x_6 = (5, 1)^T.$$

Apply k-means clustering with $k = 2$ in Euclidean setup. For cluster initialization take $z_1 = (3, 0)^T$ and $z_2 = (5, 1)^T$. Does the k-means algorithm stop at the global minimizer of the optimization problem $(\mathcal{D})$?

Exercise 5.2 (Marginal Median) Let the dissimilarity measure between $x, y \in \mathbb{R}^m$ be given by means of the *Manhattan distance*:

$$d(x, y) = \|x - y\|_1.$$

Show that the centers from the update (2) in k-means are the clusters' *marginal medians*.

Exercise 5.3 (Eigenvalues of a Stochastic Matrix) Show that the eigenvalues of a stochastic $(n \times n)$-matrix P are bounded by one, i.e. $|\lambda_r| \leq 1$ for all $r = 1, \ldots, n$.

Exercise 5.4 (Spectral Clustering) The following linkage matrix with five persons is given:

$$
W = \begin{pmatrix}
 & \boxed{1} & \boxed{2} & \boxed{3} & \boxed{4} & \boxed{5} \\
\boxed{1} & 0 & 1 & 5 & 0 & 10 \\
\boxed{2} & 1 & 0 & 8 & 0 & 0 \\
\boxed{3} & 5 & 8 & 0 & 0 & 3 \\
\boxed{4} & 0 & 0 & 0 & 0 & 12 \\
\boxed{5} & 10 & 0 & 3 & 12 & 0
\end{pmatrix}.
$$

Cluster them into two groups by means of spectral clustering.

Exercise 5.5 (Time Series Clustering) The *Pearson correlation coefficient* between two time series $x, y \in \mathbb{R}^m$ is defined as

$$\text{Pearson}(x, y) = \frac{\sigma_{xy}}{\sigma_x \cdot \sigma_y},$$

where σ_{xy} stands for the covariance of vectors x and y, whereas σ_x, σ_y stand for their standard deviations, respectively. Let the dissimilarity measure for time series be taken as

$$d(x, y) = \frac{1 - \text{Pearson}(x, y)}{2}.$$

Derive a formula for the update (2) of centers in k-means based on the Pearson correlation coefficient. Interpret the time series clustering as applied to stock prices.

Exercise 5.6 (Product Clustering) Let us represent products by binary vectors $x \in \{0, 1\}^m$, where $x_j = 1$ means that it has been consumed by the j-th customer, and $x_j = 0$ otherwise. The *Jaccard coefficient* associated with the products $x, y \in \{0, 1\}^m$ is defined as the share of customers $j \in \{1, \ldots, m\}$ who bought both products x and y, in comparison to the customers who bought just one of them:

$$J(x, y) = \frac{\#\{j \mid x_j = 1 \text{ and } y_j = 1\}}{\#\{j \mid x_j = 1 \text{ or } y_j = 1\}}.$$

Originally, Jaccard (1902) developed this "coefficient de communauté florale" within the field of botany. Let the dissimilarity measure for products be taken as

$$d(x, y) = 1 - J(x, y).$$

Apply k-means clustering with $k = 2$ based on the Jaccard coefficient to the products

$$x_1 = (0, 1, 1)^T, \quad x_2 = (1, 0, 1)^T, \quad x_3 = (0, 0, 1)^T, \quad x_4 = (1, 0, 0)^T, \quad x_5 = (1, 1, 1)^T.$$

Initialize the centers by taking $z_1 = (0, 1, 1)^T$ and $z_2 = (1, 1, 1)^T$.

Linear Regression

6

In statistics, *linear regression* is the most popular approach to modeling the relationship between an endogenous variable of response and several exogenous variables aiming to explain the former. It is crucial in linear regression to estimate unknown weights put on exogenous variables, in order to obtain the endogenous variable, from the data. The applications of linear regression just in economics are so abundant that all of them are barely to mention. To name a few, we refer to the *econometric analysis* of relationships between GDP output and unemployment rate, known as *Okun's law*, or between price and risk, known as *capital asset pricing model*. The use of linear regression is twofold. First, after fitting the linear regression it becomes possible to predict the endogenous variable by observing the exogenous variables. Second, the strength of the relationship between the endogenous and exogenous variables can be quantified. In particular, it can be clarified whether some exogenous variables may have no linear relationship with the endogenous variable at all, or identified which subsets of exogenous variables may contain redundant information about the endogenous variable. In this chapter, we discuss the meanwhile classical technique of *ordinary least squares* for linear regression. The ordinary least squares problem is derived by means of the maximum likelihood estimation, where the error terms are assumed to follow the Gauss distribution. We show that the use of the *OLS estimator* is favorable from the statistical point of view. Namely, it is a best unbiased linear estimator, as *Gauss-Markov theorem* says. From the numerical perspective we emphasize that the OLS estimator may suffer instability, especially due to possible *multicollinearity* in the data. To overcome this obstacle, the ℓ_2-regularization approach is proposed. By following the technique of maximum a posteriori estimation, we thus arrive at the *ridge regression*. Although biased, the ridge estimator reduces variance, hence, gains computational stability. Finally, we perform stability analysis of the OLS and ridge estimators in terms of the *condition number* of the underlying data matrix.

© Springer-Verlag GmbH Germany, part of Springer Nature 2021
V. Shikhman, D. Müller, *Mathematical Foundations of Big Data Analytics*,
https://doi.org/10.1007/978-3-662-62521-7_6

6.1   Motivation: Econometric Analysis

Econometric analysis is understood as an application of statistical methods to economic data in order to give empirical content to economic relationships. More precisely, it is "the quantitative analysis of actual economic phenomena based on the concurrent development of theory and observation, related by appropriate methods of inference", as pointed out by Samuelson et al. (1954). The introductory economics textbook by Samuelson and Nordhaus (2004) describes econometrics as allowing economists "to sift through mountains of data to extract simple relationships". As a typical example we mention the *Okun's law*, which relates GDP output to the unemployment rate:

$$\text{change of unemployment rate} = w_0 + w_1 \cdot \text{change of GDP output} + \varepsilon,$$

where the weights w_0, w_1 are to be statistically estimated, and ε is an error term. In Okun's original statement in 1962, the slope of $w_1 \approx -0.3$ is advocated. This means that a decrease in output of 3% leads to an increase in the unemployment rate of nearly 1%. While the Okun's law fits the data for most countries, the weight w_1 in the relationship, i.e. the effect of a one-percent change in output on the unemployment rate, varies across countries. E.g., Ball et al. (2017) estimate that it is -0.15 in Japan, -0.45 in the United States, and -0.85 in Spain. Note that the Okun's law stipulates an empirically relevant dependence between output an unemployment, rather than claims a causal relationship. Vice versa, a casual dependance between output and unemployment lays the foundation for the Okun's law. Namely, the shifts in aggregate demand cause output to fluctuate around potential. These output movements cause firms to hire and fire workers, changing employment. In turn, changes in employment move the unemployment rate in the opposite direction. In general, it is crucial to use economic reasoning for model selection, especially for deciding which variables to include in econometric analysis.

Let us present a mathematical framework for econometric analysis. For that, we are given a data set of exogenous and endogenous variables $(x_i, y_i) \in \mathbb{R}^{m-1} \times \mathbb{R}, i = 1, \ldots, n$. Let us assume that the relationship between the dependent variable y and the vector of regressors x is linear. This relationship is modeled through some error terms ε_i, $i = 1, \ldots, n$,—unobserved random variables that add noise to the linear relationship between the exogenous and endogenous variables, i.e. for $i = 1, \ldots, n$ it holds:

$$y_i = w_0 + (x_i)_1 \cdot w_1 + \ldots + (x_i)_{m-1} \cdot w_{m-1} + \varepsilon_i,$$

where $w_1, \ldots, w_{m-1} \in \mathbb{R}$ are some unknown weights and $w_0 \in \mathbb{R}$ plays the role of a bias. Here, we made use of the so-called *weak exogeneity* assumption, which essentially means that the exogenous x-variables can be treated as fixed values, rather than random variables. In other words, the exogenous variables are assumed to be error-free, that is, not contaminated with measurement errors. Shortly, the *linear regression* from above can be written in matrix form:

$$y = X \cdot w + \varepsilon, \qquad (\mathcal{LR})$$

where $y \in \mathbb{R}^n$ is the data vector of endogenous variables and ε consists of n random errors:

$$y = (y_1, \ldots, y_n)^T, \quad \varepsilon = (\varepsilon_1, \ldots, \varepsilon_n)^T.$$

The vector of weights $w \in \mathbb{R}^m$ is given by

$$w = (w_0, w_1, \ldots, w_{m-1})^T.$$

The data $(n \times m)$-matrix of exogenous variables is

$$X = \begin{pmatrix} 1 & (x_1)_1 & \cdots & (x_1)_{m-1} \\ \vdots & \vdots & \ddots & \vdots \\ 1 & (x_n)_1 & \cdots & (x_n)_{m-1} \end{pmatrix}.$$

Let us suppose that the number n of data points exceeds the number m of explanatory variables, i. e. $n > m$. This is e.g. the case if the data generation is cheap and n is relatively large. Alternatively, we may already have identified relevant explanatory variables, so that m is relatively small. Within the econometric analysis it is crucial to adjust weights w for the exogenous x-variables predict the endogenous y-variable well enough. In fact, a fitted linear regression model can be used to identify the relationship between a single exogenous variable x_j and the endogenous variable y when all the other exogenous variables in the model are held fixed. Specifically, the weight w_j can be viewed as the expected change in y if solely x_j undertakes a one-unit change. This is sometimes called the *unique effect* of x_j on y. The notion of unique effect is appealing when studying a complex systems where multiple interrelated factors influence the independent variable. In some cases, it can literally help to quantify the causal effect of an intervention that is linked to the value of an exogenous variable. However, such interpretations need to be treated with caution. In many cases multiple regression analysis fails to clarify the relationships between the exogenous and endogenous variables, in particular, when the former are correlated with each other or are not assigned following a study design.

6.2   Results

6.2.1   Ordinary Least Squares

We discuss the basic technique of ordinary least squares for linear regression in detail.

6.2.1.1 Maximum Likelihood Estimation

Let us start by specifying the error ε in $(\mathcal{LR})$. We assume that the error terms $\varepsilon_1, \ldots, \varepsilon_n$ are drawn identically and independently from the *Gauss distribution* $\mathcal{N}\left(0, \sigma^2\right)$ with zero mean and variance $\sigma^2 > 0$. In particular, we have for expectation and variance of ε:

$$\mathbb{E}(\epsilon) = 0, \quad \mathrm{Var}(\varepsilon) = \sigma^2 \cdot I,$$

where I denotes the identity matrix. A consequence of these assumptions is that the endogenous y-variable is independent across observations, conditional on the exogenous x-variables. In what follows, we eliminate from the notation the dependence on X, to make it look simpler. We emphasize that the data set $x_1, \ldots, x_n$ of exogenous x-variables is considered fixed, so all the randomness associated with y is due to the noise source ε. The Gauss probability densities of the y-variable are given as follows:

$$p\left(y_i \mid w\right) = \frac{1}{\sqrt{2\pi} \cdot \sigma} \cdot e^{-\frac{1}{2} \cdot \left(\frac{y_i - (X \cdot w)_i}{\sigma}\right)^2}, \quad i = 1, \ldots, n.$$

The conditional probability density, under the model, of observing the data is their product:

$$p\left(y \mid w\right) = \prod_{i=1}^{n} p\left(y_i \mid w\right).$$

Further, we apply *Bayes theorem* to derive the posterior distribution of the weights w:

$$p\left(w \mid y\right) = \frac{p\left(y \mid w\right) \cdot p(w)}{p(y)},$$

where $p(y)$ is the probability density of the endogenous y-variables, and $p(w)$ is a prior distribution of the weights w. Let us assume that all values of w are equally likely, i. e. the prior distribution $p(w)$ is *uniform*. Hence, in order to obtain weights, which better explain the observations, it is reasonable to maximize the so-called *likelihood function*, see e.g. Hendry and Nielsen (2014):

$$L(w) = p\left(y \mid w\right).$$

Equivalently, let us consider the *log-likelihood* instead:

$$\ln L(w) = \ln p\,(y \mid w) = \ln \prod_{i=1}^{n} p\,(y_i \mid w) = \sum_{i=1}^{n} \ln p\,(y_i \mid w)$$

$$= \sum_{i=1}^{n} \ln \frac{1}{\sqrt{2\pi} \cdot \sigma} \cdot e^{-\frac{1}{2} \cdot \left(\frac{y_i - (X \cdot w)_i}{\sigma} \right)^2}$$

$$= n \cdot \ln \frac{1}{\sqrt{2\pi} \cdot \sigma} - \frac{1}{2\sigma^2} \cdot \underbrace{\sum_{i=1}^{n} \left(y_i - (X \cdot w)_i \right)^2}_{= \|y - X \cdot w\|_2^2}.$$

The leading and the multiplicative constant can be omitted here for optimization purposes. The *maximum likelihood estimation* for adjusting weights provides then the *ordinary least squares* problem:

$$\min_{w} \; \frac{1}{2} \cdot \|y - X \cdot w\|_2^2. \tag{$\mathcal{OLS}$}$$

In other words, the weights w are chosen to minimize the regression residuum $y - X \cdot w$ with respect to the Euclidean norm $\|\cdot\|_2$.

6.2.1.2 Normal Equation

The optimization problem ($\mathcal{OLS}$) is explicitly solvable. In order to obtain its solution, we state the corresponding necessary optimality condition:

$$\nabla \left(\frac{1}{2} \cdot \|y - X \cdot w\|_2^2 \right) = 0.$$

Since we have:

$$\|y - X \cdot w\|_2^2 = (y - X \cdot w)^T \cdot (y - X \cdot w) = y^T \cdot y - 2 \cdot y^T \cdot X \cdot w + w^T \cdot X^T \cdot X \cdot w,$$

it holds for the gradient:

$$\nabla \left(\frac{1}{2} \cdot \|y - X \cdot w\|_2^2 \right) = -X^T \cdot y + X^T \cdot X \cdot w.$$

Hence, the necessary optimality condition for ($\mathcal{OLS}$) yields the *normal equation*:

$$X^T \cdot X \cdot w = X^T \cdot y.$$

If the $(m \times m)$-matrix $X^T \cdot X$ regular, its unique solution is called the *OLS estimator*:

$$w_{OLS} = \left(X^T \cdot X\right)^{-1} \cdot X^T \cdot y.$$

Due to convexity, the OLS estimator w_{OLS} solves the optimization problem $(\mathcal{OLS})$. For the matrix $X^T \cdot X$ to be regular, it is sufficient to require that the data $(n \times m)$-matrix X is of full rank, i.e.

$$\mathrm{rank}(X) = m.$$

In fact, the application of Exercise 3.6 ensures that the product $X^T \cdot X$ of the $(m \times n)$-matrix X^T and the $(n \times m)$-matrix X is of rank m, provided both of them are of full rank m too. Throughout the remaining part of the chapter we assume that the matrix X is of full rank m.

6.2.1.3 Pseudoinverse
It is convenient to rewrite the formula for the OLS estimator as

$$w_{OLS} = X^\dagger \cdot y,$$

where we use the *pseudoinverse* of X by setting:

$$X^\dagger = \left(X^T \cdot X\right)^{-1} \cdot X^T.$$

The pseudoinverse deserves its name, since it is a left inverse of X, but not necessarily a right one. In fact, it holds:

$$X^\dagger \cdot X = \left(X^T \cdot X\right)^{-1} \cdot X^T \cdot X = I.$$

Useful will be also a representation of $X^\dagger$ in terms of the singular value decomposition of X. For that, we write in reduced form:

$$X = U \cdot \Sigma \cdot V,$$

where the columns of an $(n \times m)$-matrix U and the rows of an $(m \times m)$-matrix V are orthogonal:

$$U^T \cdot U = I, \quad V \cdot V^T = I,$$

and the diagonal $(m \times m)$-matrix Σ has positive singular values $\sigma_j(X)$, $j = 1, \dots, m$, on its main diagonal. Let us substitute the singular value decomposition of X into the pseudoinverse:

$$X^\dagger = \left((U \cdot \Sigma \cdot V)^T \cdot U \cdot \Sigma \cdot V \right)^{-1} \cdot (U \cdot \Sigma \cdot V)^T$$

$$= \left(V^T \cdot \Sigma \cdot \underbrace{U^T \cdot U}_{=I} \cdot \Sigma \cdot V \right)^{-1} \cdot V^T \cdot \Sigma \cdot U^T$$

$$= \left(V^T \cdot \Sigma^2 \cdot V \right)^{-1} \cdot V^T \cdot \Sigma \cdot U^T$$

$$= V \cdot \Sigma^{-2} \cdot \underbrace{V \cdot V^T}_{=I} \cdot \Sigma \cdot U^T = V^T \cdot \Sigma^{-1} \cdot U^T,$$

where we used the fact that the quadratic matrix V is orthogonal, i. e. $V^{-1} = V^T$. We have thus shown that a singular value decomposition of the pseudoinverse is

$$X^\dagger = V^T \cdot \Sigma^{-1} \cdot U^T.$$

Hence, the singular values of $X^\dagger$ are reciprocal to those of X, i.e. for $j = 1, \dots, m$ it holds:

$$\sigma_j \left(X^\dagger \right) = \frac{1}{\sigma_j(X)}.$$

In particular, for the largest and smallest singular values we have:

$$\sigma_{max} \left(X^\dagger \right) = \frac{1}{\sigma_{min}(X)}, \quad \sigma_{min} \left(X^\dagger \right) = \frac{1}{\sigma_{max}(X)}.$$

6.2.1.4 OLS Estimator

We point out that the OLS estimator inherits randomness, since it depends on the error ε:

$$w_{OLS} = X^\dagger \cdot y = X^\dagger \cdot (X \cdot w + \varepsilon) = \underbrace{X^\dagger \cdot X}_{=I} \cdot w + X^\dagger \cdot \varepsilon = w + X^\dagger \cdot \varepsilon,$$

where the weights w play now the role of the true model parameter. Recalling that the error ε has zero mean, we compute the expectation of w_{OLS}:

$$\mathbb{E}\left(w_{OLS}\right) = \mathbb{E}\left(w + X^{\dagger} \cdot \varepsilon\right) = \underbrace{\mathbb{E}\left(w\right)}_{=w} + X^{\dagger} \cdot \underbrace{\mathbb{E}\left(\varepsilon\right)}_{=0} = w.$$

From here we see that the OLS estimator w_{OLS} recovers the true weights w on average, i.e. it is *unbiased*. Let us compute its variance:

$$\mathrm{Var}\left(w_{OLS}\right) = \mathrm{Var}\left(w + X^{\dagger} \cdot \varepsilon\right) = \mathrm{Var}\left(X^{\dagger} \cdot \varepsilon\right) = X^{\dagger} \cdot \underbrace{\mathrm{Var}(\varepsilon)}_{=\sigma^2 \cdot I} \cdot \left(X^{\dagger}\right)^{T}$$

$$= \sigma^2 \cdot X^{\dagger} \cdot \left(X^{\dagger}\right)^{T} = \sigma^2 \cdot \left(X^{T} \cdot X\right)^{-1},$$

since it holds for the last step:

$$X^{\dagger} \cdot \left(X^{\dagger}\right)^{T} = X^{\dagger} \cdot \left(\left(X^{T} \cdot X\right)^{-1} \cdot X^{T}\right)^{T} = \underbrace{X^{\dagger} \cdot X}_{=I} \cdot \left(X^{T} \cdot X\right)^{-1} = \left(X^{T} \cdot X\right)^{-1}.$$

It turns out that the OLS estimator has the lowest variance within the class of linear unbiased estimators. Loosely speaking, the mistake while estimating the true parameter w is minimal if using the OLS estimator w_{OLS}. The precise statement of this fact is known as Gauss-Markov theorem.

6.2.1.5 Gauss-Markov Theorem

The assumptions of Gauss-Markov theorem concern the random error ε in $(\mathcal{LR})$:

GM1: *Strict exogeneity* means that the errors in the regression should have zero mean, i.e. for all $i = 1, \ldots, n$ it holds:

$$\mathbb{E}\left(\varepsilon_i\right) = 0.$$

GM2: *Homoscedasticity* states that different values of the endogenous variable have the same variance in their errors, regardless of the values of the exogenous variables, i.e. for all $i = 1, \ldots, n$ it holds:

$$\mathrm{Var}\left(\varepsilon_i\right) = \sigma^2.$$

GM3: *Independence* implies that the errors of the endogenous variables are uncorrelated with each other, i.e. for all $i, j = 1, \ldots, n$ and $i \neq j$ it holds:

$$\mathrm{Cor}\left(\varepsilon_i, \varepsilon_j\right) = 0.$$

For short, we say that the error terms $\varepsilon_1, \ldots, \varepsilon_n$ are uncorrelated with zero mean and the same variance $\sigma^2 > 0$. Note that they do not need to follow the Gaussian distribution anymore. As above, we equivalently have for expectation and variance of ε:

$$\mathbb{E}(\epsilon) = 0, \quad \mathrm{Var}(\varepsilon) = \sigma^2 \cdot I.$$

Further, we consider *linear estimators* of the form:

$$w_{lin} = C \cdot y,$$

where an $(m \times n)$-matrix C is not allowed to depend on the unobservable true weights w, but does rather depend on the observable data matrix X. Suppose now that the linear estimator w_{lin} is unbiased, i.e.

$$\mathbb{E}\left(w_{lin}\right) = w.$$

Let us compute the variance of w_{lin}. For that, we use the following representation:

$$C = X^\dagger + D,$$

where the $(m \times n)$-matrix D is appropriately chosen. Then, we have:

$$w_{lin} = C \cdot y = \left(X^\dagger + D\right) \cdot (X \cdot w + \varepsilon)$$

$$= \underbrace{X^\dagger \cdot X}_{=I} \cdot w + D \cdot X \cdot w + \left(X^\dagger + D\right) \cdot \varepsilon = w + D \cdot X \cdot w + \left(X^\dagger + D\right) \cdot \varepsilon.$$

It holds for the expectation of w_{lin}:

$$\mathbb{E}\left(w_{lin}\right) = \mathbb{E}\left(w + D \cdot X \cdot w\right) + \left(X^\dagger + D\right) \cdot \underbrace{\mathbb{E}(\varepsilon)}_{=0} = w + D \cdot X \cdot w.$$

Since the estimator w_{lin} is unbiased, and the matrix D does not depend on w, we deduce:

$$D \cdot X = 0.$$

In particular, it follows:

$$D \cdot \left(X^{\dagger}\right)^{T} = \underbrace{D \cdot X}_{=0} \cdot \left(X^{T} \cdot X\right)^{-1} = 0.$$

Finally, we get for the variance of w_{lin}:

$$\text{Var}\left(w_{lin}\right) = \text{Var}\left(w + D \cdot X \cdot w + \left(X^{\dagger} + D\right) \cdot \varepsilon\right)$$

$$= \text{Var}\left(\left(X^{\dagger} + D\right) \cdot \varepsilon\right) = \left(X^{\dagger} + D\right) \cdot \underbrace{\text{Var}(\varepsilon)}_{=\sigma^{2} \cdot I} \cdot \left(X^{\dagger} + D\right)^{T}$$

$$= \sigma^{2} \cdot \underbrace{X^{\dagger} \cdot \left(X^{\dagger}\right)^{T}}_{=(X^{T} \cdot X)^{-1}} + \sigma^{2} \cdot \underbrace{X^{\dagger} \cdot D^{T}}_{=0} + \sigma^{2} \cdot \underbrace{D \cdot \left(X^{\dagger}\right)^{T}}_{=0} + \sigma^{2} \cdot D \cdot D^{T}$$

$$= \sigma^{2} \cdot \left(X^{T} \cdot X\right)^{-1} + \sigma^{2} \cdot D \cdot D^{T}.$$

Overall, the variance of w_{lin} exceeds that of w_{OLS} by a positive semidefinite matrix $\sigma^{2} \cdot D \cdot D^{T}$:

$$\text{Var}\left(w_{lin}\right) = \text{Var}\left(w_{OLS}\right) + \sigma^{2} \cdot D \cdot D^{T}.$$

Under the Gauss-Markov assumptions (i)–(iii), w_{OLS} is thus proven to be a *best linear unbiased estimator* (BLUE). Meaning, that among all linear unbiased estimators w_{lin} the OLS estimator w_{OLS} is one with the minimal variance and, therefore, most efficient. This conclusion of *Gauss-Markov theorem* highlights the importance of the OLS estimator from the statistical point of view.

6.2.1.6 Multicollinearity
Now, we turn our attention to the regularity assumption we put on the matrix $X^{T} \cdot X$ in order to derive the OLS estimator w_{OLS}. For that, we required that the data $(n \times m)$-matrix X is of full rank, i.e.

$$\text{rank}(X) = m.$$

In this case, we say that the linear regression lacks *multicollinearity* in predictors. Multicollinearity can be triggered by having two or more correlated exogenous variables. In some sense, multicollinear variables contain the same information about the endogenous variable. If nominally different variables actually quantify the same phenomenon then they are redundant. Alternatively, if the variables get different names and perhaps employ different numeric measurement scales, but are highly correlated with each other, then they suffer from redundancy. In practice, we rarely face perfect multicollinearity in a data set. More commonly, there is an approximate linear relationship among two or more exogenous variables. A standard measure of such linear dependence of the columns of X is the *condition number*:

$$\kappa(X) = \frac{\sigma_{max}(X)}{\sigma_{min}(X)},$$

where $\sigma_{max}(X)$ and $\sigma_{min}(X)$ are the largest and the smallest singular values of the $(m \times n)$-matrix X, respectively. We recall that, due to the full rank of X, all its m singular values are positive, in particular, $\sigma_{max}, \sigma_{min} > 0$. However, as soon as the columns of X become linearly dependent, i.e. the rank of X is getting less than m, at least one of its singular values vanishes, thus, the condition number of X explodes. This motivates to use $\kappa(X)$ as a measure for multicollinearity of the data. A linear regression with a high condition number of X is said to be *ill-conditioned*.

6.2.1.7 Stability

One of the negative effects of multicollinearity is that small changes of the data can lead to large changes in the regression model, even resulting in changes of weights' signs. Let us illustrate this instability phenomenon by studying of how much the OLS estimator w_{OLS} changes if the endogenous y-variable is due to some measurement inaccuracies. To model the latter, we assume that instead of y we observe $\widehat{y}$, and consider the corresponding OLS estimator:

$$\widehat{w}_{OLS} = X^\dagger \cdot \widehat{y}.$$

Let us compare the relative error in OLS estimators with that of measurement inaccuracies:

$$\frac{\|w_{OLS} - \widehat{w}_{OLS}\|_2}{\|w_{OLS}\|_2} : \frac{\|y - \widehat{y}\|_2}{\|y\|_2} = \frac{\|X^\dagger \cdot (y - \widehat{y})\|_2}{\|y - \widehat{y}\|_2} \cdot \frac{\|y\|_2}{\|X^\dagger \cdot y\|}.$$

Our goal is now to bound both terms on the right from above by means of the condition number $\kappa(X)$. For the first term, we set $y - \widehat{y} = z$ and obtain due to homogeneity, cf. Exercise 4.3:

$$\frac{\|X^\dagger \cdot (y - \widehat{y})\|_2}{\|y - \widehat{y}\|_2} \leq \max_{z \in \mathbb{R}^n} \frac{\|X^\dagger \cdot z\|_2}{\|z\|_2} = \max_{\|z\|_2 = 1} \|X^\dagger \cdot z\|_2.$$

Exercise 3.4 provides that the latter expression equals to the largest singular value of $X^\dagger$, i.e.

$$\max_{\|z\|_2=1} \left\| X^\dagger \cdot z \right\|_2 = \sigma_{max}\left(X^\dagger\right).$$

Analogously, for the second term we have:

$$\frac{\left\| X^\dagger \cdot y \right\|}{\|y\|_2} \geq \min_{z \in \mathbb{R}^n} \frac{\left\| X^\dagger \cdot z \right\|_2}{\|z\|_2} = \min_{\|z\|_2=1} \left\| X^\dagger \cdot z \right\|_2 = \sigma_{min}\left(X^\dagger\right).$$

Altogether, we obtain:

$$\frac{\left\| X^\dagger \cdot (y - \widehat{y}) \right\|_2}{\|y - \widehat{y}\|_2} \cdot \frac{\|y\|_2}{\left\| X^\dagger \cdot y \right\|} \leq \frac{\sigma_{max}\left(X^\dagger\right)}{\sigma_{min}\left(X^\dagger\right)}.$$

Recalling the relation between the singular values of X and $X^\dagger$, it holds for the upper bound:

$$\kappa\left(X^\dagger\right) = \frac{\sigma_{max}\left(X^\dagger\right)}{\sigma_{min}\left(X^\dagger\right)} = \frac{1}{\sigma_{min}(X)} : \frac{1}{\sigma_{max}(X)} = \frac{\sigma_{max}(X)}{\sigma_{min}(X)} = \kappa(X).$$

Finally, the relative error in OLS estimators with respect to measurement inaccuracies is bounded by the condition number of X:

$$\frac{\|w_{OLS} - \widehat{w}_{OLS}\|_2}{\|w_{OLS}\|_2} : \frac{\|y - \widehat{y}\|_2}{\|y\|_2} \leq \kappa(X).$$

Intuitively, if we face an ill-conditioned linear regression with a high condition number $\kappa(X)$, then possible measurement inaccuracies in endogenous y-variables may have a dramatic impact on the OLS estimator w_{OLS}.

6.2.2   Ridge Regression

In order to overcome the instability of the OLS estimator, an ℓ_2-regularization technique can be applied. The latter leads to the so-called ridge regression, as we shall see in a moment.

6.2.2.1 Maximum A Posteriori Estimation

Let us take the probabilistic perspective on the ℓ_2-regularization of $(\mathcal{LR})$ via the Euclidean norm. This can be done by applying the technique of maximum a posteriori estimation.

For that, we assume again that the error terms $\varepsilon_1, \ldots, \varepsilon_n$ are drawn identically and independently from the *Gauss distribution* $\mathcal{N}\left(0, \sigma^2\right)$ with zero mean and variance $\sigma^2 > 0$. Recall that the Gauss probability densities of the y-variable are given as follows:

$$p\left(y_i \mid w\right) = \frac{1}{\sqrt{2\pi} \cdot \sigma} \cdot e^{-\frac{1}{2} \cdot \left(\frac{y_i - (X \cdot w)_i}{\sigma}\right)^2}, \quad i = 1, \ldots, n.$$

The conditional probability density, under the model, of observing the data is their product:

$$p\left(y \mid w\right) = \prod_{i=1}^{n} p\left(y_i \mid w\right).$$

Additionally, we assume that the weights $w = (w_0, \ldots, w_{m-1})^T$ are drawn identically and independently also from the *Gauss distribution* $\mathcal{N}\left(0, \tau^2\right)$ with zero mean and variance $\tau^2 > 0$. The Gauss probability densities of the weights are given as follows:

$$p\left(w_j\right) = \frac{1}{\sqrt{2\pi} \cdot \tau} \cdot e^{-\frac{1}{2} \cdot \left(\frac{w_j}{\tau}\right)^2}, \quad j = 0, \ldots, m - 1.$$

Their joint probability density is the product:

$$p(w) = \prod_{j=0}^{m-1} p\left(w_j\right).$$

Further, we interpret $p(w)$ as a prior distribution, and apply *Bayes theorem* to derive the posterior distribution of the weights w:

$$p\left(w \mid y\right) = \frac{p\left(y \mid w\right) \cdot p(w)}{p(y)},$$

where $p(y)$ is the probability density of the endogenous y-variables. In order to obtain weights, which better explain the observed endogenous variables, it is reasonable to choose the *mode* of the posterior distribution. Since its denominator is always positive and does not depend on w, we may equivalently maximize the numerator:

$$N(w) = p\left(y \mid w\right) \cdot p(w).$$

This leads to the technique of *maximum a posteriori estimation*, see e.g. Murphy (2012):

$$\max_{w} N(w). \qquad\qquad (\mathcal{MAP})$$

In order to simplify $(\mathcal{MAP})$, it is convenient to maximize the logarithm of the numerator instead:

$$\ln N(w) = \ln p\,(y \mid w) \cdot p(w) = \ln p\,(y \mid w) + \ln p(w)$$

$$= \sum_{i=1}^{n} \ln \frac{1}{\sqrt{2\pi} \cdot \sigma} \cdot e^{-\frac{1}{2} \cdot \left(\frac{y_i - (X \cdot w)_i}{\sigma} \right)^2} + \sum_{j=0}^{m-1} \ln \frac{1}{\sqrt{2\pi} \cdot \tau} \cdot e^{-\frac{1}{2} \cdot \left(\frac{w_j}{\tau} \right)^2}$$

$$= n \cdot \ln \frac{1}{\sqrt{2\pi} \cdot \sigma} - \frac{1}{2} \cdot \sum_{i=1}^{n} \left(\frac{y_i - (X \cdot w)_i}{\sigma} \right)^2 + m \cdot \ln \frac{1}{\sqrt{2\pi} \cdot \tau} - \frac{1}{2} \cdot \sum_{j=0}^{m-1} \left(\frac{w_j}{\tau} \right)^2$$

$$= n \cdot \ln \frac{1}{\sqrt{2\pi} \cdot \sigma} + m \cdot \ln \frac{1}{\sqrt{2\pi} \cdot \tau} - \frac{1}{2\sigma^2} \cdot \|y - X \cdot w\|_2^2 - \frac{1}{2\tau^2} \cdot \|w\|_2^2 \,.$$

After omitting the leading constant and appropriately scaling the remaining terms, $(\mathcal{MAP})$ is equivalent to the following optimization problem:

$$\min_{w} \frac{1}{2} \cdot \|y - X \cdot w\|_2^2 + \frac{\sigma^2}{2\tau^2} \cdot \|w\|_2^2 \,.$$

By setting $\lambda = \frac{\sigma^2}{\tau^2}$, we finally obtain the *ridge regression*:

$$\min_{w} \frac{1}{2} \cdot \|y - X \cdot w\|_2^2 + \frac{\lambda}{2} \cdot \|w\|_2^2 \,. \tag{$\mathcal{Ridge}$}$$

Ridge regression is often referred to as *Tikhonov regularization*, suggested by Tikhonov and Arsenin (1977) for treating ill-posed inverse problems. It is particularly useful to mitigate the problem of multicollinearity in linear regression, which commonly occurs in models with large numbers of parameters. Note that the choice of the prior Gauss distribution for the weights w induces the ℓ_2-regularization term $\frac{\lambda}{2} \cdot \|w\|_2^2$. Its role is to induce penalty on the weights. The positive Tikhonov parameter λ regularizes the weights in the sense that, if they take large values, the objective function in ridge regression is penalized. In other words, ridge regression shrinks the weights and helps to reduce the model complexity and multicollinearity.

6.2.2.2 Ridge Estimator

The optimization problem $(\mathcal{Ridge})$ is explicitly solvable. In order to obtain its solution, we state the corresponding necessary optimality condition:

$$\nabla \left(\frac{1}{2} \cdot \|y - X \cdot w\|_2^2 + \frac{\lambda}{2} \cdot \|w\|_2^2 \right) = 0 \,.$$

It holds for the gradient:

$$\nabla \left(\frac{1}{2} \cdot \| y - X \cdot w \|_2^2 + \frac{\lambda}{2} \cdot \| w \|_2^2 \right) = -X^T \cdot y + X^T \cdot X \cdot w + \lambda \cdot w.$$

Hence, the necessary optimality condition for ($\mathcal{R}idge$) yields the regularized normal equation:

$$\left(X^T \cdot X + \lambda \cdot I \right) \cdot w = X^T \cdot y.$$

Note that the $(m \times m)$-matrix $X^T \cdot X + \lambda \cdot I$ is positive definite. In fact, for all $\xi \in \mathbb{R}^m$ with $\xi \neq 0$ it easily follows:

$$\xi^T \cdot \left(X^T \cdot X + \lambda \cdot I \right) \cdot \xi = \xi^T \cdot X^T \cdot X \cdot \xi + \lambda \cdot \xi^T \cdot \xi = \underbrace{\| X \cdot \xi \|_2^2}_{\geq 0} + \underbrace{\lambda \cdot \| \xi \|_2^2}_{> 0} > 0.$$

Consequently, the matrix $X^T \cdot X + \lambda \cdot I$ is regular, and that independently of the full rank assumption on X. Even in case of multicollinearity—if the rank of X is less than m—we obtain the unique solution of the regularized normal equation, called *ridge estimator*:

$$w_{ridge} = \left(X^T \cdot X + \lambda \cdot I \right)^{-1} \cdot X^T \cdot y.$$

Due to convexity, the ridge estimator w_{ridge} solves the optimization problem ($\mathcal{R}idge$). As above we rewrite the formula for the ridge estimator as

$$w_{ridge} = X_\lambda^\dagger \cdot y,$$

where we use the regularized pseudoinverse of X by setting:

$$X_\lambda^\dagger = \left(X^T \cdot X + \lambda \cdot I \right)^{-1} \cdot X^T.$$

6.2.2.3 Condition Number

Let us demonstrates the effect of the Tikhonov parameter λ on the condition number of the regularized problem. For that, we first derive a singular value decomposition of $X_\lambda^\dagger$ by substituting $X = U \cdot \Sigma \cdot V$ into the regularized pseudoinverse:

$$X_\lambda^\dagger = \left((U \cdot \Sigma \cdot V)^T \cdot U \cdot \Sigma \cdot V + \lambda \cdot I \right)^{-1} \cdot (U \cdot \Sigma \cdot V)^T$$

$$= \left(V^T \cdot \Sigma \cdot \underbrace{U^T \cdot U}_{=I} \cdot \Sigma \cdot V + \lambda \cdot I \right)^{-1} \cdot V^T \cdot \Sigma \cdot U^T$$

$$= \left(V^T \cdot \left(\Sigma^2 + \lambda \cdot I \right) \cdot V \right)^{-1} \cdot V^T \cdot \Sigma \cdot U^T$$

$$= V^T \cdot \left(\Sigma^2 + \lambda \cdot I \right)^{-1} \cdot \underbrace{V \cdot V^T}_{=I} \cdot \Sigma \cdot U^T$$

$$= V^T \cdot \left(\Sigma^2 + \lambda \cdot I \right)^{-1} \cdot \Sigma \cdot U^T,$$

where we again used the fact that the quadratic matrix V is orthogonal, i.e. $V^{-1} = V^T$. We have thus shown that a singular value decomposition of the regularized pseudoinverse is

$$X_\lambda^\dagger = V^T \cdot \left(\Sigma^2 + \lambda \cdot I \right)^{-1} \cdot \Sigma \cdot U^T.$$

Hence, the singular values of $X_\lambda^\dagger$ can be related to those of X, and we get for $j = 1, \ldots, m$:

$$\sigma_j \left(X_\lambda^\dagger \right) = \frac{\sigma_j(X)}{\sigma_j^2(X) + \lambda}.$$

From here we see that the singular values of the regularized pseudoinverse are shrinked:

$$\sigma_j \left(X_\lambda^\dagger \right) = \frac{\sigma_j(X)}{\sigma_j^2(X) + \lambda} \leq \frac{1}{\sigma_j(X)} = \sigma_j \left(X^\dagger \right).$$

We further get access to the largest and smallest singular values of $X_\lambda^\dagger$ by means of the function:

$$f(t) = \frac{t}{t^2 + \lambda}.$$

It is easy to see that f attains its maximum at $\sqrt{\lambda}$. Moreover, it is strictly increasing for $t < \sqrt{\lambda}$, and strictly decreasing for $t > \sqrt{\lambda}$, see Fig. 6.1. Therefore, we have:

$$\sigma_{max} \left(X_\lambda^\dagger \right) = \max_{j=1,\ldots,m} \frac{\sigma_j(X)}{\sigma_j^2(X) + \lambda} \leq \frac{\sqrt{\lambda}}{\sqrt{\lambda}^2 + \lambda} = \frac{1}{2\sqrt{\lambda}},$$

$$\sigma_{min} \left(X_\lambda^\dagger \right) = \min_{j=1,\ldots,m} \frac{\sigma_j(X)}{\sigma_j^2(X) + \lambda} = \min \left\{ \frac{\sigma_{min}(X)}{\sigma_{min}^2(X) + \lambda}, \frac{\sigma_{max}(X)}{\sigma_{max}^2(X) + \lambda} \right\}.$$

Fig. 6.1 Graph of
$f(t) = \frac{t}{t^2 + \lambda}$

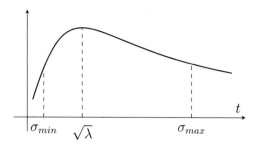

The condition number of $X_\lambda^\dagger$ can be thus estimated as follows:

$$\kappa\left(X_\lambda^\dagger\right) = \frac{\sigma_{max}\left(X_\lambda^\dagger\right)}{\sigma_{min}\left(X_\lambda^\dagger\right)} \le \frac{1}{2\sqrt{\lambda}} : \min\left\{\frac{\sigma_{min}(X)}{\sigma_{min}^2(X) + \lambda}, \frac{\sigma_{max}(X)}{\sigma_{max}^2(X) + \lambda}\right\}$$

$$= \frac{1}{2\sqrt{\lambda}} \cdot \max\left\{\frac{\sigma_{min}^2(X) + \lambda}{\sigma_{min}(X)}, \frac{\sigma_{max}^2(X) + \lambda}{\sigma_{max}(X)}\right\}$$

$$= \frac{1}{2} \cdot \max\left\{\frac{\sigma_{min}(X)}{\sqrt{\lambda}} + \frac{\sqrt{\lambda}}{\sigma_{min}(X)}, \frac{\sigma_{max}(X)}{\sqrt{\lambda}} + \frac{\sqrt{\lambda}}{\sigma_{max}(X)}\right\}.$$

By specifying the Tikhonov parameter $\lambda = \sigma_{min} \cdot \sigma_{max}$, we finally obtain:

$$\kappa\left(X_\lambda^\dagger\right) \le \frac{1}{2} \cdot \left(\sqrt{\frac{1}{\kappa(X)}} + \sqrt{\kappa(X)}\right).$$

We conclude that the condition number of $X_\lambda^\dagger$ is bounded by the *square root* of the condition number of X. This fact suggests that the ridge estimator w_{ridge} is much more stable with respect to measurement inaccuracies in endogenous y-variables than it is the case for the OLS estimator w_{OLS}.

6.2.2.4 Bias-Variance Tradeoff

As we have just seen, the ridge estimator has favorable stability properties in comparison to the OLS estimator. However, there is a price to pay. In contrast to the OLS estimator, the ridge estimator is biased. To see this, we assume for simplicity that the columns of the data matrix X are not only linearly independent, but even pairwise orthogonal:

$$X^T \cdot X = n \cdot I.$$

Then, the formula for the ridge estimator reads:

$$w_{ridge} = X_\lambda^\dagger \cdot y = \left(\underbrace{X^T \cdot X}_{=n \cdot I} + \lambda \cdot I \right)^{-1} \cdot X^T \cdot (X \cdot w + \varepsilon) = \frac{n}{n + \lambda} \cdot w + \frac{1}{n + \lambda} \cdot X^T \cdot \varepsilon.$$

In expectation, w_{ridge} does not provide the true weights w:

$$\mathbb{E}\left(w_{ridge}\right) = \frac{n}{n + \lambda} \cdot \underbrace{\mathbb{E}\left(w\right)}_{=w} + \frac{1}{n + \lambda} \cdot X^T \cdot \underbrace{\mathbb{E}\left(\varepsilon\right)}_{=0} = \frac{n}{n + \lambda} \cdot w \neq w.$$

In other words, the ridge estimator is *biased*. What can be said about its variance? We obtain:

$$\text{Var}\left(w_{ridge}\right) = \text{Var}\left(\frac{n}{n + \lambda} \cdot w + \frac{1}{n + \lambda} \cdot X^T \cdot \varepsilon \right) = \frac{1}{(n + \lambda)^2} \cdot \text{Var}\left(X^T \cdot \varepsilon\right)$$

$$= \frac{1}{(n + \lambda)^2} \cdot X^T \cdot \underbrace{\text{Var}(\varepsilon)}_{=\sigma^2 \cdot I} \cdot \left(X^T\right)^T = \frac{1}{(n + \lambda)^2} \cdot \sigma^2 \cdot \underbrace{X^T \cdot X}_{=n \cdot I}$$

$$= \frac{n}{(n + \lambda)^2} \cdot \sigma^2 \cdot \underbrace{I}_{=n \cdot (X^T \cdot X)^{-1}} = \frac{n^2}{(n + \lambda)^2} \cdot \sigma^2 \cdot \underbrace{\left(X^T \cdot X\right)^{-1}}_{=\text{Var}(w_{OLS})}$$

$$= \frac{n^2}{(n + \lambda)^2} \cdot \text{Var}\left(w_{OLS}\right).$$

Since $\frac{n^2}{(n+\lambda)^2} < 1$, we conclude that the variance of w_{OLS} exceeds that of w_{ridge} in the sense that the matrix difference $\text{Var}\left(w_{OLS}\right) - \text{Var}\left(w_{ridge}\right)$ is positive definite. This fact reveals the bias-variance tradeoff in statistical estimation. The *bias-variance tradeoff* says that estimators with a lower bias have a higher variance, and vice versa, cf. Exercise 6.6. The ℓ_2-regularization introduces bias into the regression model, but reduces variance as compared with the OLS estimator. Although the OLS estimator provides unbiased weights and demonstrates superior performance on average, the ridge estimator is more robust with respect to measurement inaccuracies.

6.3   Case Study: Capital Asset Pricing

In finance, the capital asset pricing model describes the relationship between systematic risk and expected return for assets. It is widely used for pricing risky securities and generating expected returns for assets given the risk of those assets and the cost of capital.

The capital asset pricing model, introduced by Sharpe (1964), is based on the earlier work on the modern portfolio theory due to Markowitz (1952). Let us approach the former by means of linear regression. For that, we denote by r the return on a traded capital asset, whereas r^M stands for the return on the whole stock market, say, expressed by means of a market index as e.g. Dow Jones. The fixed return on a risk-free asset is r^F, which could correspond e. g. to U.S. Government bonds. The *capital asset pricing model* assumes linear dependence between those random variables:

$$r - r^F = \beta \cdot \left(r^M - r^F \right), \qquad\qquad (\mathcal{CAPM})$$

where $\beta \in \mathbb{R}$ quantifies how much the return on the asset changes with respect to a change in the return on the market.

Task 1 Show that the asset *risk premium* is proportional to the market risk premium, i.e.

$$\mathbb{E}(r) - r^F = \beta \cdot \left(\mathbb{E}\left(r^M \right) - r^F \right).$$

Hint 1 Take expectation in $(\mathcal{CAPM})$.

Task 2 Show that beta measures the *systematic risk* or, in other words, the market related risk of an asset, i.e.

$$\beta = \frac{\mathrm{Cov}\left(r, r^M \right)}{\mathrm{Var}\left(r^M \right)}.$$

Hint 2 Using $(\mathcal{CAPM})$ and Task 1, it follows:

$$r - \mathbb{E}(r) = r^F + \beta \cdot \left(r^M - r^F \right) - r^F - \beta \cdot \left(\mathbb{E}\left(r^M \right) - r^F \right) = \beta \cdot \left(r^M - \mathbb{E}\left(r^M \right) \right).$$

Then, we obtain:

$$\mathrm{Cov}\left(r, r^M \right) = \mathbb{E}\left((r - \mathbb{E}(r)) \cdot \left(r^M - \mathbb{E}\left(r^M \right) \right) \right) = \mathbb{E}\left(\beta \cdot \left(r^M - \mathbb{E}\left(r^M \right) \right)^2 \right) = \beta \cdot \mathrm{Var}\left(r^M \right).$$

Task 3 Given a data set of returns $\left(r_i, r_i^M \right) \in \mathbb{R} \times \mathbb{R}$, $i = 1, \ldots, n$, estimate the unknown weights α and β by means of the linear regression for $(\mathcal{CAPM})$:

$$r_i - r^F = \alpha + \beta \cdot \left(r_i^M - r^F \right) + \varepsilon_i,$$

where the error terms ε_i, $i = 1, \ldots, n$, are drawn identically and independently from the Gauss distribution $\mathcal{N}\left(0, \sigma^2\right)$ with zero mean and variance $\sigma^2 > 0$.

Hint 3 The OLS estimator provides:

$$\alpha = \frac{\sum_{i=1}^{n} \left(r_i - r^F\right) - \beta \cdot \sum_{i=1}^{n} \left(r_i^M - r^F\right)}{n},$$

$$\beta = \frac{n \cdot \sum_{i=1}^{n} \left(r_i^M - r^F\right) \cdot \left(r_i - r^F\right) - \sum_{i=1}^{n} \left(r_i^M - r^F\right) \cdot \sum_{i=1}^{n} \left(r_i - r^F\right)}{n \cdot \sum_{i=1}^{n} \left(r_i^M - r^F\right)^2 - \left(\sum_{i=1}^{n} \left(r_i^M - r^F\right)\right)^2}.$$

Task 4 Suppose that an asset is purchased at price p and later sold at price q. By using the estimated beta, derive and interpret a formula for the price of an asset:

$$p = \frac{\mathbb{E}(q)}{1 + r^F + \beta \cdot \left(\mathbb{E}\left(r^M\right) - r^F\right)}.$$

Hint 4 Use $r = \frac{q-p}{p}$ and Task 1. The asset price p is discounted by means of the risk-adjusted interest rate $r^F + \beta \cdot \left(\mathbb{E}\left(r^M\right) - r^F\right)$.

Task 5 Suppose that an asset is purchased at price p and later sold at price q. Derive and interpret the certainty equivalent pricing formula:

$$p = \frac{1}{1 + r^F} \cdot \left(\mathbb{E}(q) + \frac{\mathrm{Cov}\left(q, r^M\right) \cdot \left(\mathbb{E}\left(r^M\right) - r^F\right)}{\mathrm{Var}\left(r^M\right)}\right).$$

Hint 5 Use Tasks 2, 4, and 5. The term in the brackets can be viewed as the risk-adjusted expectation of q, alias its *certainty equivalent*.

6.4   Exercises

Exercise 6.1 (Okun's Law) Examine the Okun's law on the following panel data for Germany:

Year	GDP output in billion USD	Unemployment in %
1991	1.86	5.317
1992	2.12	6.323
1993	2.07	7.675
1994	2.21	8.728
1995	2.59	8.158
1996	2.50	8.825
1997	2.22	9.863
1998	2.24	9.788
1999	2.20	8.855
2000	1.95	7.917
2001	1.95	7.773
2002	2.01	8.482
2003	2.51	9.779
2004	2.82	10.727
2005	2.86	11.167
2006	3.00	10.250
2007	3.44	8.658
2008	3.75	7.524
2009	3.42	7.742
2010	3.42	6.966
2011	3.76	5.824
2012	3.54	5.379
2013	3.75	5.231
2014	3.88	4.981
2015	3.36	4.624
2016	3.47	4.122

Exercise 6.2 (Pseudoinverse) Let an $(n \times m)$-matrix X be of full rank m. Show that the following *Moore-Penrose conditions* for its pseudoinverse $X^\dagger$ are valid:

(i) $X \cdot X^\dagger \cdot X = X$,

(ii) $X^\dagger \cdot X \cdot X^\dagger = X^\dagger$,

(iii) $\left(X \cdot X^\dagger \right)^T = X \cdot X^\dagger$.

Exercise 6.3 (Hilbert Matrix) *Hilbert $(n \times n)$-matrices* are defined as follows:

$$
H_n = \left(\frac{1}{i+j-1}\right) = \begin{pmatrix} 1 & 1/2 & 1/3 & \cdots & 1/n \\ 1/2 & 1/3 & & & \\ 1/3 & & & & \\ \vdots & & & & \vdots \\ 1/n & & & \cdots & 1/2n-1 \end{pmatrix}.
$$

(i) Compute the condition numbers $\kappa\,(H_n)$ for $n = 5, 10, 15$ numerically. Compare your calculations with the asymptotic result for large n:

$$
\kappa\,(H_n) \sim \frac{(1+\sqrt{2})^{4n}}{\sqrt{n}}.
$$

(ii) Solve the linear system of equations $H_n \cdot x = y$ numerically by taking $y = H_n \cdot e$ for $n = 5, 10, 15$. Do you obtain the solution $x = e$, why not?

Exercise 6.4 (Vandermonde Matrix) The rows of the *Vandermonde $(m \times m)$-matrix* consist of the powers of given numbers $\alpha_1, \ldots, \alpha_m \in \mathbb{R}$:

$$
V = \begin{pmatrix} 1 & \alpha_1 & \alpha_1^2 & \cdots & \alpha_1^{m-1} \\ 1 & \alpha_2 & \alpha_2^2 & \cdots & \alpha_2^{m-1} \\ \vdots & \vdots & \vdots & \ddots & \vdots \\ 1 & \alpha_m & \alpha_m^2 & \cdots & \alpha_m^{m-1} \end{pmatrix}.
$$

Show by induction that the determinant of a Vandermonde matrix can be represented as

$$
\det(V) = \prod_{1 \le i < j \le m} (\alpha_j - \alpha_i).
$$

Exercise 6.5 (Polynomial Regression) Let an endogenous y-variable depend on an exogenous x-variable in polynomial manner, i.e.

$$
y = w_0 + x \cdot w_1 + x^2 \cdot w_2 + \ldots + x^{m-1} \cdot w_{m-1} + \varepsilon,
$$

where $w_0, w_1, \ldots, w_{m-1} \in \mathbb{R}$ are unknown coefficients, and the random error ε follows the Gauss distribution $\mathcal{N}\,(0, \sigma^2)$ with zero mean and variance $\sigma^2 > 0$. For a given data set $(x_i, y_i) \in \mathbb{R} \times \mathbb{R}$, $i = 1, \ldots, n$, reformulate the *polynomial regression* as an ordinary least squares problem. Assume that $n > m$, and all x_i, $i = 1, \ldots, n$, are pairwise different. Provide the unique solution of the corresponding normal equation.

Exercise 6.6 (Mean Squared Error) In order to predict true weights w, we consider the linear estimator $w_{lin} = C \cdot y$ with an $(m \times n)$-matrix C and $y = X \cdot w + \varepsilon$, where random errors ε fulfill Gauss-Markov conditions GM1-GM3. Show the bias-variance decomposition of the *mean squared error*:

$$\mathbb{E} \, \|w_{lin} - w\|_2^2 = \|\mathbb{E} \, (w_{lin}) - w\|_2^2 + \text{trace} \, (\text{Var} \, (w_{lin})) \, .$$

By assuming $X^T \cdot X = n \cdot I$, compute and compare the mean squared errors for the OLS estimator w_{OLS} and the ridge estimator w_{ridge}. Discuss the corresponding bias-variance tradeoff. Compute the Tikhonov parameter λ, which minimizes the mean square error of the ridge estimator w_{ridge}.

Sparse Recovery

7

With the increasing amount of information available, the cost of processing high-dimensional data becomes a critical issue. In order to reduce the model complexity, the concept of sparsity is widely used over last decades. *Sparsity* refers in this context to the requirement that most of the model parameters are equal or close to zero. As a relevant application, we mention the *variable selection* from econometrics, where zero entries correspond to irrelevant features. Another important application concerns the *compressed sensing* from signal processing, where just a few linear measurements are usually enough in order to decode sparse signals. In this chapter, the sparsity of a vector will be measured with respect to the *zero norm*. By using the latter, we state the problem of determining the sparsest vector subject to linear equality constraints. Its unique solvability in terms of *spark* of the constraint matrix is given. For gaining convexity, we substitute the zero norm by the Manhattan norm. The corresponding regularized optimization problem is known as *basis pursuit*. We show that the basis pursuit admits sparse solutions under the null space property of the underlying matrix. Further, a probabilistic technique of maximum a posteriori estimation is applied to derive the *least absolute shrinkage and selection operator*. This optimization problem is similar to the basis pursuit, but allows the application of efficient numerical schemes from convex optimization. We discuss the *iterative shrinkage-thresholding algorithm* for its solution.

7.1 Motivation: Variable Selection

In data science, *variable selection* is the process of identifying a subset of important variables for use in model construction. The central premise when using a variable selection technique is that the data contain some features that are either redundant or irrelevant, and can thus be removed without incurring much loss of information. Variable

© Springer-Verlag GmbH Germany, part of Springer Nature 2021
V. Shikhman, D. Müller, *Mathematical Foundations of Big Data Analytics*,
https://doi.org/10.1007/978-3-662-62521-7_7

Fig. 7.1 Hughes phenomenon

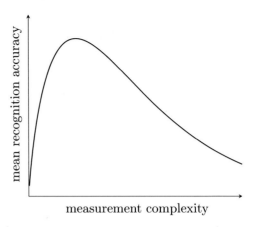

selection techniques are used for two main reasons. First, the simplification of models naturally makes them easier to interpret by users. It is usually not only required to suggest a reliable model, but also to reduce the number of independent variables therein. In famous words of Albert Einstein: "Make everything as simple as possible, but not simpler". Second, by means of variable selection the curse of dimensionality can be avoided. The *curse of dimensionality* refers to difficulties that arise when analyzing and organizing data in high-dimensional spaces. In big data analytics an enormous amount of data is typically required to ensure that there are several samples with each combination of values. A rule of thumb says that there should be at least five training examples for each dimension in the representation. With a fixed number of data samples, the predictive power of a regressor first increases, as the number of features grows, but then it decreases. This *peaking phenomenon* has been first observed by Hughes (1968) in the framework of pattern recognition, see Fig. 7.1. Variable selection tries to overcome the *Hughes phenomenon* by reducing the number of features.

Let us present the problem of variable selection in mathematical terms. For that, we are given a data set of exogenous and endogenous variables $(x_i, y_i) \in \mathbb{R}^{m-1} \times \mathbb{R}, i = 1, \ldots, n$. We assume that the endogenous y-variable depends linearly on the exogenous x-variables, i. e. for $i = 1, \ldots, n$ it holds:

$$y_i = w_0 + (x_i)_1 \cdot w_1 + \ldots + (x_i)_{m-1} \cdot w_{m-1},$$

where $w_1, \ldots, w_{m-1} \in \mathbb{R}$ are some unknown weights and $w_0 \in \mathbb{R}$ plays the role of a bias. Often, these n equations of *linear regression* are stacked together and written in matrix form as a system:

$$y = X \cdot w, \qquad\qquad (\mathcal{SLE})$$

where $y \in \mathbb{R}^n$ is the data vector of endogenous variables and $w \in \mathbb{R}^m$ is the vector of weights:

$$y = (y_1, \ldots, y_n)^T, \quad w = (w_0, w_1, \ldots, w_{m-1})^T.$$

The data $(n \times m)$-matrix of exogenous variables is

$$X = \begin{pmatrix} 1 & (x_1)_1 & \cdots & (x_1)_{m-1} \\ \vdots & \vdots & \ddots & \vdots \\ 1 & (x_n)_1 & \cdots & (x_n)_{m-1} \end{pmatrix}.$$

Let us suppose that the number m of explanatory variables considerably exceeds the number n of data points, i.e. $n < m$. This is e.g. the case if the data generation is costly, and n is relatively small. Alternatively, we may not be able to a priori identify explanatory variables, but rather want to select the most important ones a posteriori, and m is relatively large. As a consequence, the system $(\mathcal{SLE})$ of linear equations will be underdetermined, and, in case of solvability, will admit multiple solutions. We are interested in those solutions, which are sparse, i.e. with the least number of non-zero entries. The *sparsity* of a vector $w \in \mathbb{R}^m$ can be measured by the so-called *zero norm*, which counts the number of its non-zero elements:

$$\|w\|_0 = \#\{j \mid w_j \neq 0\}.$$

In contrast to a norm, the zero norm is not absolutely homogeneous. Nevertheless, it is positive definite and satisfies the triangle inequality. For $\alpha \in \mathbb{R}$ and $v, w \in \mathbb{R}^m$ it holds namely, see Exercise 7.1:

- Positive definite: $\|w\|_0 = 0$ if and only if $w = 0$.
- Failure of absolute homogeneity: $\|\alpha \cdot w\|_0 = |\alpha| \cdot \|w\|_0$ if and only if $\alpha \in \{0, \pm 1\}$ or $w = 0$.
- Triangle inequality: $\|v + w\|_0 \leq \|v\|_0 + \|w\|_0$.

By using the zero norm, we face the following problem of variable selection:

$$\min_w \|w\|_0 \quad \text{s.t.} \quad y = X \cdot w. \qquad (\mathcal{P}_0)$$

Note that, given a solution w of $(\mathcal{P}_0)$, we have identified a few explanatory variables which are consistent with the linear regression model $(\mathcal{SLE})$. They correspond to the indices from the support of w:

$$\text{supp}(w) = \{j \mid w_j \neq 0\}.$$

Thus, by solving the optimization problem ($\mathcal{P}_0$), variable selection can be successively performed.

7.2   Results

7.2.1   Lasso Regression

Our goal is—apart from analyzing sparsity patterns in $\mathcal{P}_0$—to derive the lasso regression, standing for least absolute shrinkage and selection operator. This is necessary in order to convexify the optimization problem $\mathcal{P}_0$.

7.2.1.1 Spark

Let us first state a sufficient condition for a feasible vector w to solve ($\mathcal{P}_0$). How is it possible to decide just by looking at the sparsity pattern of w if it is optimal? To address this issue, we introduce the notion of spark of a matrix. For the *spark* of an $(n \times m)$-matrix X we have:

$$\mathrm{spark}(X) = \text{minimal number of linearly dependent columns of } X.$$

Note that, although spark and rank are in some ways similar, they are totally different. The rank of a matrix X is defined as the maximal number of its columns that are linearly independent. Spark, on the other hand, is the minimal number of columns from X that are linearly dependent. In general it holds:

$$\mathrm{spark}(X) \leq \mathrm{rank}(X) + 1.$$

We claim that for w to uniquely solve ($\mathcal{P}_0$) it is sufficient to fulfill:

$$\|w\|_0 < \frac{\mathrm{spark}(X)}{2}.$$

To see this, let v be an arbitrary feasible point of ($\mathcal{P}_0$) distinct from w. The difference $v - w$ lies in the nullspace of X, since it holds:

$$X \cdot (v - w) = X \cdot v - X \cdot w = y - y = 0.$$

Hence, the columns corresponding to the non-zero entries of $v - w$ are linearly dependent. In terms of spark, we have:

$$\mathrm{spark}(X) \leq \|v - w\|_0.$$

The triangle inequality for the zero norm implies further:

$$\text{spark}(X) \leq \|v - w\|_0 \leq \|v\|_0 + \|w\|_0.$$

Then, we have by the derived inequality:

$$\|v\|_0 \geq \text{spark}(X) - \|w\|_0 > \text{spark}(X) - \frac{\text{spark}(X)}{2} = \frac{\text{spark}(X)}{2}.$$

Hence, the number of non-zero entries of v is greater than that of w. As a consequence, v cannot be optimal, and we are done. Overall, we conclude that, for w to uniquely solve $(\mathcal{P}_0)$, it is sufficient that its zero norm is less than the half of the spark of X. Unfortunately, the computation of $\text{spark}(X)$ requires to examine the linear dependence of column subsets of cardinality up to $\text{rank}(X) + 1$. In worst case with $\text{rank}(X) = n$, this combinatorial process is of huge complexity:

$$\sum_{k=1}^{n+1} \binom{m}{k}.$$

It is not surprising therefore that the optimization problem $(\mathcal{P}_0)$ turns out to be *NP-hard*, and can barely be solved efficiently, see e. g. Tillmann and Pfetsch (2013).

7.2.1.2 Basis Pursuit

In order to tackle $(\mathcal{P}_0)$ numerically, it is common to substitute the zero norm $\| \cdot \|_0$ by the Manhattan norm $\| \cdot \|_1$, see Chen et al. (1998). This ℓ_1-*regularization* approach is based on the fact that the arising optimization problem, called *basis pursuit*, is convex:

$$\min_{w} \|w\|_1 \quad \text{s. t.} \quad y = X \cdot w. \tag{$\mathcal{P}_1$}$$

Intuitively, it is rather likely that solutions of the basis pursuit $(\mathcal{P}_1)$ are sparse, see Fig. 7.2. In fact, the blue contours of the Manhattan norm are rhombus-shaped. Thus, the intersection point of the optimal contour with the feasible set, being a linear subspace, will lie on the axes. This advantageous behavior of the Manhattan norm automatically induces sparsity while solving $(\mathcal{P}_1)$. Note that e.g. the Euclidean norm $\| \cdot \|_2$ is not suitable for this purpose. The red contours of the Euclidean norm are circle-shaped. Since they will merely touch the feasible set at the optimal points, and the latter would not normally generate any zero entries.

7.2.1.3 Null Space Property

Now, let us turn our attention to the relation between the optimization problems $(\mathcal{P}_0)$ and $(\mathcal{P}_1)$. For that, let w be a solution of $(\mathcal{P}_0)$ with s non-zero entries, i. e. $\|w\|_0 = s$. Is it also a solution of $(\mathcal{P}_1)$? Yes, if the matrix X has the *null space property* of order s.

Fig. 7.2 Basis pursuit

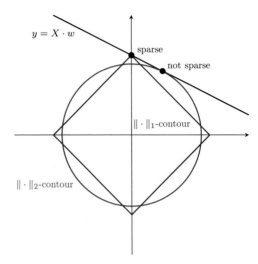

Loosely speaking, the latter says that any s components of a vector from the nullspace of X are strictly dominated by the others with respect to the Manhattan norm. To put it in mathematical terms, for all nontrivial vectors $u \in \text{null}(X)$ with $u \neq 0$, and subsets S with $|S| = s$ it should hold:

$$\|u_S\|_1 < \|u_{S^c}\|_1 \,,$$

where u_S and u_{S^c} arise from u by setting its components to zero, which correspond to the subsets S^c and S, respectively. Since sparsity bound s is usually taken small, the size of S is much less than of its complement S^c. Thus, the inequality in the null space property seems to be a quite reasonable assumption. Further, for the solution w of $(\mathcal{P}_0)$ we set $S = \text{supp}(w)$, and obtain for any feasible v:

$$\|w\|_1 = \|w_S\|_1 = \|w_S - v_S + v_S\|_1 \leq \|w_S - v_S\|_1 + \|v_S\|_1$$

$$= \|(w - v)_S\|_1 + \|v_S\|_1 = \|u_S\|_1 + \|v_S\|_1 \,,$$

where we denoted the difference of vectors by $u = w - v$. Since $u \in \text{null}(X)$ and $|S| = s$, we proceed due to the null space property of X:

$$\|u_S\|_1 + \|v_S\|_1 < \|u_{S^c}\|_1 + \|v_S\|_1 = \|(w - v)_{S^c}\|_1 + \|v_S\|_1$$

$$= \|w_{S^c} - v_{S^c}\|_1 + \|v_S\|_1 = \|0 - v_{S^c}\|_1 + \|v_S\|_1 = \|v\|_1 \,.$$

Overall, the Manhattan norm of any feasible v exceeds that of w:

$$\|w\|_1 < \|v\|_1 \,.$$

Hence, w does not only solve $(\mathcal{P}_0)$, but also $(\mathcal{P}_1)$. This conclusion justifies that the convex optimization problem $(\mathcal{P}_1)$ can be tried instead of dealing with the NP-hard combinatorial optimization problem $(\mathcal{P}_0)$. At least, if X satisfies the null space property of order s, and $(\mathcal{P}_0)$ admits an s-sparse solution, we shall recover the latter by tackling $(\mathcal{P}_1)$.

7.2.1.4 Maximum A Posteriori Estimation

Let us take the probabilistic perspective on the ℓ_1-regularization of $(\mathcal{P}_0)$. This can be done by applying the technique of maximum a posteriori estimation. For that, we write the linear regression between the endogenous y-variable and exogenous x-variables by means of the error terms $\varepsilon = (\varepsilon_1, \ldots, \varepsilon_n)^T$, i.e. for all $i = 1, \ldots, n$ it holds:

$$y_i = w_0 + (x_i)_1 \cdot w_1 + \ldots + (x_i)_{m-1} \cdot w_{m-1} + \varepsilon_i,$$

or, equivalently:

$$y = X \cdot w + \varepsilon.$$

We assume that the error terms $\varepsilon_1, \ldots, \varepsilon_n$ are drawn identically and independently from the *Gauss distribution* $\mathcal{N}(0, \sigma^2)$ with zero mean and variance $\sigma^2 > 0$. A consequence of these assumptions is that the endogenous y-variable is independent across observations, conditional on the exogenous x-variables. In what follows, we eliminate from the notation the dependence on X, to make it look simpler. We emphasize that the data set $x_1, \ldots, x_n$ of exogenous x-variables is considered fixed, so all the randomness associated with y is due to the noise source ε. The Gauss probability densities of the y-variable are given as follows:

$$p(y_i \mid w) = \frac{1}{\sqrt{2\pi} \cdot \sigma} \cdot e^{-\frac{1}{2} \cdot \left(\frac{y_i - (X \cdot w)_i}{\sigma} \right)^2}, \quad i = 1, \ldots, n.$$

The conditional probability density, under the model, of observing the data is their product:

$$p(y \mid w) = \prod_{i=1}^{n} p(y_i \mid w).$$

Additionally, we assume that the weights $w = (w_0, \ldots, w_{m-1})^T$ are drawn identically and independently from the *Laplace distribution* $\mathcal{L}(0, \tau)$ with zero location and scaling parameter $\tau > 0$, see Exercise 7.4. The Laplace probability densities of the weights are given as follows:

$$p(w_j) = \frac{1}{2\tau} \cdot e^{-\frac{|w_j|}{\tau}}, \quad j = 0, \ldots, m-1.$$

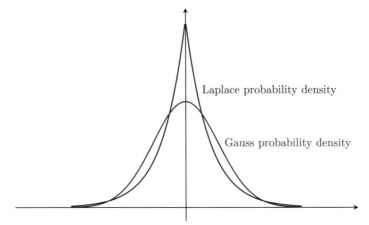

Fig. 7.3 Laplace vs. Gauss distribution

Their joint probability density is the product:

$$p(w) = \prod_{j=0}^{m-1} p\left(w_j\right).$$

The choice of the Laplace distribution induces sparsity of weights. Although the probability density of the Laplace distribution is reminiscent of the Gauss distribution, however, whereas the latter is expressed in terms of the squared difference, the Laplace density is expressed in terms of the absolute difference. Consequently, the Laplace distribution has fatter tails than the normal distribution. In other words, the weights, which follow the Laplace distribution, are more likely to be close to the zero mean, see Fig. 7.3.

Further, we interpret $p(w)$ as a prior distribution, and apply *Bayes theorem* to derive the posterior distribution of the weights w:

$$p(w \mid y) = \frac{p(y \mid w) \cdot p(w)}{p(y)},$$

where $p(y)$ is the probability density of the endogenous y-variables. In order to obtain weights, which better explain the observed endogenous variables, it is reasonable to choose the *mode* of the posterior distribution. Since its denominator is always positive and does not depend on w, we may equivalently maximize the numerator:

$$N(w) = p(y \mid w) \cdot p(w).$$

This leads to the technique of *maximum a posteriori estimation*, see e.g. Murphy (2012):

$$\max_{w} N(w). \qquad\qquad (\mathcal{MAP})$$

In order to simplify ($\mathcal{MAP}$), it is convenient to maximize the logarithm of the numerator instead:

$$\ln N(w) = \ln p\,(y \mid w) \cdot p(w) = \ln p\,(y \mid w) + \ln p(w)$$

$$= \sum_{i=1}^{n} \ln \frac{1}{\sqrt{2\pi} \cdot \sigma} \cdot e^{-\frac{1}{2} \left(\frac{y_i - (X \cdot w)_i}{\sigma} \right)^2} + \sum_{j=0}^{m-1} \ln \frac{1}{2\tau} \cdot e^{-\frac{|w_j|}{\tau}}$$

$$= n \cdot \ln \frac{1}{\sqrt{2\pi} \cdot \sigma} - \frac{1}{2} \cdot \sum_{i=1}^{n} \left(\frac{y_i - (X \cdot w)_i}{\sigma} \right)^2 + m \cdot \ln \frac{1}{2\tau} - \sum_{j=0}^{m-1} \frac{|w_j|}{\tau}$$

$$= n \cdot \ln \frac{1}{\sqrt{2\pi} \cdot \sigma} + m \cdot \ln \frac{1}{2\tau} - \frac{1}{2\sigma^2} \cdot \|y - X \cdot w\|_2^2 - \frac{1}{\tau} \cdot \|w\|_1 \,.$$

After omitting the leading constant and appropriately scaling the remaining terms, ($\mathcal{MAP}$) is equivalent to the following optimization problem:

$$\min_{w} \; \frac{1}{2} \cdot \|y - X \cdot w\|_2^2 + \frac{\sigma^2}{\tau} \cdot \|w\|_1 \,.$$

By setting $\lambda = \frac{\sigma^2}{\tau}$, we finally obtain the *least absolute shrinkage and selection operator*:

$$\min_{w} \; \frac{1}{2} \cdot \|y - X \cdot w\|_2^2 + \lambda \cdot \|w\|_1 \,. \qquad (\mathcal{Lasso})$$

A variant of ($\mathcal{Lasso}$) has been first introduced by Tibshirani (1996) in the context of least squares, see Exercise 7.5. Let us comment on the relation of ($\mathcal{Lasso}$) to the optimization problem ($\mathcal{P}_1$). In ($\mathcal{P}_1$), the goal is to find the sparsest weights among those which perfectly explain the underlying linear regression. In ($\mathcal{Lasso}$), we pursue the combination of these two objectives by introducing the balancing parameter λ. The price to pay is that the linear regression is relaxed and its residual with respect to the Euclidean norm is now minimized.

7.2.2 Iterative Shrinkage-Thresholding Algorithm

The optimization problem ($\mathcal{Lasso}$) is convex, thus, allowing the application of efficient numerical schemes. For that, we put it into the framework of *composite convex optimization*:

$$\min_{w} \; F(w) = \underbrace{\frac{1}{2} \cdot \|y - X \cdot w\|_2^2 + \lambda \cdot \|w\|_1}_{=f(w)} \,.$$

Note that the objective function in ($\mathcal{L}asso$) is composed of two parts: both f and $\|\cdot\|_1$ are convex functions, but whereas f is twice continuously differentiable, the Manhattan norm $\|\cdot\|_1$ is nonsmooth. The idea of composite convex optimization is to construct a quadratic overestimation of the smooth part, and to use the fact that the nonsmooth part is relatively simple. This will lead us to the optimization technique called proximal gradient descent and, in particular, to the iterative shrinkage-thresholding algorithm.

7.2.2.1 Quadratic Overestimation

First, let us assume that f has *L-Lipschitz continuous gradients* with respect to the Euclidean norm $\|\cdot\|_2$, i.e. for all $u, v \in \mathbb{R}^m$ it holds:

$$\|\nabla f(u) - \nabla f(v)\|_2 \leq L \cdot \|u - v\|_2.$$

By the fundamental theorem of calculus and the chain rule, we have:

$$f(w) - f(v) = \int_0^1 f'(v + s \cdot (w - v))\, \mathrm{d}s = \int_0^1 \nabla^T f(v + s \cdot (w - v)) \cdot (w - v)\, \mathrm{d}s.$$

The Cauchy-Schwarz inequality, cf. Exercise 2.2, and the Lipschitz continuity provide by setting $u = v + s \cdot (w - v)$:

$$\left(\nabla^T f(u) - \nabla^T f(v)\right) \cdot (w - v) \leq \|\nabla f(u) - \nabla f(v)\|_2 \cdot \|w - v\|_2$$

$$\leq L \cdot \|u - v\|_2 \cdot \|w - v\|_2$$

$$= L \cdot s \cdot \|w - v\|_2^2.$$

Altogether, we obtain:

$$f(w) - f(v) - \nabla^T f(v) \cdot (w - v) \leq \int_0^1 L \cdot s \cdot \|w - v\|_2^2\, \mathrm{d}s = \frac{L}{2} \cdot \|w - v\|_2^2.$$

Thus, we get the *quadratic overestimation* of f at a given vector $v \in \mathbb{R}^m$:

$$f(w) \leq \underbrace{f(v) + \nabla^T f(v) \cdot (w - v) + \frac{L}{2} \cdot \|w - v\|_2^2}_{= \tilde{f}(w, v)}$$

In particular, for $v = w$ we have equality:

$$\tilde{f}(w, w) = f(w) + \nabla^T f(w) \cdot (w - w) + \frac{L}{2} \cdot \|w - w\|_2^2 = f(w).$$

From the quadration overestimation, we also see that the *Lipschitz constant L* measures the curvature of the function f. Thus, it can be characterized by the upper bound on the Hesse matrix of f, i.e. for all $w \in \mathbb{R}^m$ and $\xi \in \mathbb{R}^m$ it holds:

$$\xi^T \cdot \nabla^2 f(w) \cdot \xi \leq L \cdot \|\xi\|_2^2.$$

By computing the Hesse matrix of f:

$$\nabla^2 f(w) = \nabla^2 \left(\frac{1}{2} \cdot \|y - X \cdot w\|_2^2 \right) = X^T \cdot X,$$

we need to determine the Lipschitz constant as

$$L = \max_\xi \frac{\xi^T \cdot \nabla^2 f(v) \cdot \xi}{\|\xi\|_2^2} = \max_{\|\xi\|_2 = 1} \xi^T \cdot X^T \cdot X \cdot \xi = \max_{\|\xi\|_2 = 1} \|X \cdot \xi\|_2^2.$$

Here, we encounter the *spectral norm* of the matrix X defined as

$$\|X\|_{2,2} = \max_{\|\xi\|_2 = 1} \|X \cdot \xi\|_2.$$

As we have seen in Exercise 3.4, the spectral norm of a matrix equals to its largest singular value $\sigma_{max}(X)$. Finally, we have:

$$L = \sigma_{max}^2(X).$$

7.2.2.2 Soft-Thresholding

Next, let us substitute the objective function in ($\mathcal{L}asso$) by its quadratic overestimation at a given vector $v \in \mathbb{R}^m$:

$$\min_w \tilde{f}(w, v) + \lambda \cdot \|w\|_1. \tag{$\mathcal{QO}$}$$

The unique solution of ($\mathcal{QO}$) can be given explicitly. To derive it, we first recall that

$$\tilde{f}(w, v) = f(v) + \nabla^T f(v) \cdot (w - v) + \frac{L}{2} \cdot \|w - v\|_2^2.$$

By ignoring those terms, which do not depend on the variable w, ($\mathcal{QO}$) reduces to

$$\min_w \sum_{j=0}^{m-1} \nabla_j^T f(v) \cdot w_j + \frac{L}{2} \cdot (w_j - v_j)^2 + \lambda \cdot |w_j|.$$

The latter objective function is separable. Thus, we can equivalently solve m one-dimensional optimization problems for every $j = 0, \ldots, m - 1$:

$$\min_{w_j} \nabla_j^T f(v) \cdot w_j + \frac{L}{2} \cdot (w_j - v_j)^2 + \lambda \cdot |w_j|.$$

Let us consider the following cases:

(a) If $w_j > 0$, then the necessary optimality condition reads:

$$\nabla_j^T f(v) + L \cdot (w_j - v_j) + \lambda = 0.$$

Form here, we obtain:

$$w_j = \left[v_j - \frac{\nabla_j^T f(v)}{L} \right] - \frac{\lambda}{L}.$$

(b) If $w_j < 0$, then the necessary optimality condition reads:

$$\nabla_j^T f(v) + L \cdot (w_j - v_j) - \lambda = 0.$$

Form here, we obtain:

$$w_j = \left[v_j - \frac{\nabla_j^T f(v)}{L} \right] + \frac{\lambda}{L}.$$

Depending on the sign of the derived w_j, case (a) or (b) occurs, otherwise the optimal w_j vanishes. Overall, for the solution thus holds:

$$w_j = \begin{cases} \left[v_j - \frac{\nabla_j^T f(v)}{L} \right] - \frac{\lambda}{L}, & \text{if } \left[v_j - \frac{\nabla_j^T f(v)}{L} \right] > \frac{\lambda}{L}, \\[2mm] 0, & \text{if } -\frac{\lambda}{L} \leq \left[v_j - \frac{\nabla_j^T f(v)}{L} \right] \leq \frac{\lambda}{L}, \\[2mm] \left[v_j - \frac{\nabla_j^T f(v)}{L} \right] + \frac{\lambda}{L}, & \text{if } \left[v_j - \frac{\nabla_j^T f(v)}{L} \right] < -\frac{\lambda}{L}. \end{cases}$$

Alternatively, the solution can be expressed by

$$w_j = \left[\left| v_j - \frac{\nabla_j^T f(v)}{L} \right| - \frac{\lambda}{L} \right]_+ \cdot \text{sign}\left(v_j - \frac{\nabla_j^T f(v)}{L} \right),$$

Fig. 7.4 Soft-thresholding operator

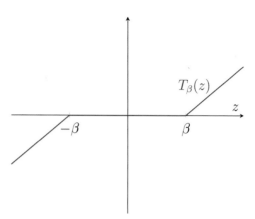

where for $a \in \mathbb{R}$ its positive part is denoted by a_+ and its sign by $\text{sign}(a)$. It is convenient to rewrite this formula by using the soft-thresholding operator. Given a threshold $\beta > 0$, the *soft-thresholding* operator shrinks the variable $z \in \mathbb{R}$ as follows, see Fig. 7.4:

$$T_\beta(z) = [|z| - \beta]_+ \cdot \text{sign}(z).$$

By means of soft-thresholding, we write in vector form:

$$w = T_{\frac{\lambda}{L}}\left(v - \frac{1}{L} \cdot \nabla f(v)\right).$$

7.2.2.3 Proximal Gradient Descent

Now, we are ready to formulate the iteration scheme for solving ($\mathcal{L}asso$). To get the next iterate $w(t + 1)$, we minimize the quadratic overestimation at the previous iterate $w(t)$. For $t = 1, 2, \ldots$ we thus set:

$$w(t + 1) = \arg\min_w \ \widetilde{f}(w, w(t)) + \lambda \cdot \|w\|_1.$$

As we have seen, this formula reduces to

$$w(t + 1) = T_{\frac{\lambda}{L}}\left(w(t) - \frac{1}{L} \cdot \nabla f(w(t))\right), \qquad (\mathcal{PGD})$$

where the soft-thresholding operator applies to the usual gradient descent with the constant stepsize $1/L$. This method is therefore called *proximal gradient descent*. Similar to the gradient descent, the values of the ($\mathcal{L}asso$) objective function do not increase during the iteration process:

$$F(w(t+1)) = f(w(t+1)) + g(w(t+1)) \leq \tilde{f}(w(t+1), w(t)) + g(w(t+1))$$

$$= \min_w \tilde{f}(w, w(t)) + g(w) \leq \tilde{f}(w(t), w(t)) + g(w(t))$$

$$= f(w(t)) + g(w(t)) = F(w(t)).$$

Moreover, the speed of convergence can be deduced, see e.g. Beck and Teboulle (2009). For any solution w^* of ($\mathcal{L}asso$) it hods:

$$F(w(t+1)) - F\left(w^*\right) \leq \frac{\lambda \cdot L \cdot \|w(1) - w^*\|_2^2}{2(t+1)}.$$

To obtain ($\mathcal{PGD}$) in explicit form, it remains to compute the gradient of f:

$$\nabla f(w) = \nabla \left(\frac{1}{2} \cdot \|y - X \cdot w\|_2^2 \right) = X^T \cdot (X \cdot w - y).$$

Finally, the general step of the *iterative shrinkage-thresholding algorithm* is

$$w(t+1) = T_{\frac{\lambda}{L}} \left(w(t) - \frac{1}{L} \cdot X^T \cdot (X \cdot w(t) - y) \right). \qquad (\mathcal{ISTA})$$

This update, widely used in the context of signal processing, induces sparsity by the shrinkage procedure. If the prediction of gradient descent is small enough, ($\mathcal{ISTA}$) sets the corresponding weight to zero, thus, automatically generating patterns of sparsity. For the convergence rate of ($\mathcal{ISTA}$) we have:

$$F(w(t+1)) - F\left(w^*\right) \leq \frac{\lambda \cdot \sigma_{max}^2(X) \cdot \|w(1) - w^*\|_2^2}{2(t+1)}.$$

It is of order $1/t+1$, which can be even improved by applying the so-called *fast iterative shrinkage-thresholding algorithm*, see again Beck and Teboulle (2009).

7.3   Case Study: Compressed Sensing

In the engineering field of *signal processing*, it is crucial to reconstruct a signal from a series of measurements. In general, this task is impossible because there is no way to reconstruct a signal during the times that the signal is not measured. Nevertheless, with prior knowledge or assumptions about the signal, e.g. the latter to be sparse, such reconstruction turns out to succeed. In mathematical terms, this corresponds to the problem of *compressed sensing* which can be stated as follows, see e.g. Foucart and Rauhut (2013). For a signal $x \in \mathbb{R}^m$, we observe n linear measurements $y_1, \ldots, y_n \in \mathbb{R}$. By means of some sensing vectors $a_1, \ldots, a_n \in \mathbb{R}^m$, it thus holds:

$$y_i = a_i^T \cdot x, \quad i = 1, \ldots, n.$$

For brevity, we define the sensing $(n \times m)$-matrix A, and the vector $y \in \mathbb{R}^n$ of linear measurements:

$$A = (a_1, \ldots, a_n)^T, \quad y = (y_1, \ldots, y_n)^T.$$

In matrix form we equivalently have:

$$y = A \cdot x.$$

Here, we assume that the number m of signal's components considerably exceeds the number n of observed measurements, i.e. $n < m$. In this case, the system $y = A \cdot x$ of linear equations is underdetermined, and, without additional information, it is impossible to recover x from y. Nevertheless, we organize the linear measurement process to do so by assuming that the signals are k-sparse. A signal $x \in \mathbb{R}^m$ is called k-sparse if it has at most k non-zero entries. For the latter, we write $x \in \Sigma_k$, where

$$\Sigma_k = \left\{ x \in \mathbb{R}^m \mid \|x\|_0 \leq k \right\}.$$

The main goal of compressed sensing is to construct a sensing matrix A, so that at least sparse signals can be *decoded* by observing corresponding linear measurements. This means that for any $y \in \mathbb{R}^n$ there should exist the unique k-sparse signal $x \in \Sigma_k$, such that $y = A \cdot x$ holds. Of course, we would like to perform as few measurements as possible for that. Hence, another task of compressed sensing is to obtain a lower bound on n, so that the decoding process is still reliable. In what follows, we first characterize sensing matrices with the decoding guarantee, and then construct a particular sensing matrix which will do the job.

Task 1 Show that $y = A \cdot x$ admits the unique solution $x \in \Sigma_k$ for any y if and only if the null space of A does not contain any nontrivial $2k$-sparse vectors, i.e.

$$\Sigma_{2k} \cap \mathrm{null}(A) = \{0\}.$$

Hint 1 Let x and z be k-sparse with $y = Ax = Az$. Then, $x - z$ is $2k$-sparse and $A \cdot (x - z) = 0$. Thus, $x - z \in \Sigma_{2k} \cap \mathrm{null}(A) = \{0\}$, and $x = z$ is uniquely decoded. Vice versa, let $v \in \Sigma_{2k} \cap \text{null}(A)$ be nontrivial. Then, there exists $x, z \in \Sigma_k$ with $\text{supp}(x) \cap \text{supp}(z)$ and $v = x - z$. Due to $A \cdot v = 0$, we have $Ax = Az$, hence, the decoding fails.

Task 2 Show that $y = A \cdot x$ admits the unique solution $x \in \Sigma_k$ for any y if and only if the matrix A is $2k$-regular, i. e. any $2k$ columns of A are linearly independent.

Hint 2 Use Task 1 and observe that for a $2k$-sparse vector $x \neq 0$, we have $A \cdot x = 0$ if and only if $2k$ columns of A, corresponding to the support of x, are linearly dependent.

Task 3 Show that the number of measurements n, needed to decode k-sparse signals, always satisfies $n \geq 2k$, i. e. at least twice as many measurements are required.

Hint 3 In virtue of Task 2, we have $\text{rank}(A) \geq 2k$. Additionally, it holds:

$$\text{rank}(A) \leq \min\{m, n\} = n.$$

Task 4 For given numbers $\alpha_1, \ldots, \alpha_n \in \mathbb{R}$, we define the so-called *Vandermonde matrix*, whose i-th column consists of the powers of α_i, $i = 1, \ldots, n$:

$$V = \begin{pmatrix} 1 & 1 & \cdots & 1 \\ \alpha_1 & \alpha_2 & \cdots & \alpha_n \\ \alpha_1^2 & \alpha_2^2 & \cdots & \alpha_n^2 \\ \vdots & \vdots & \ddots & \vdots \\ \alpha_1^{n-1} & \alpha_2^{n-1} & \cdots & \alpha_n^{n-1} \end{pmatrix}.$$

Show that V is regular if and only if $\alpha_1, \ldots, \alpha_n$ are pairwise different.

Hint 4 For the determinant of the Vandermonde matrix see Exercise 6.4:

$$\det(V) = \prod_{1 \leq i < j \leq n} (\alpha_j - \alpha_i).$$

Task 5 Let $n = 2k$ and $\alpha_1, \ldots, \alpha_m \in \mathbb{R}$ be pairwise different. Show that the following sensing $(n \times m)$-matrix guarantees the decoding of k-sparse signals:

$$A = \begin{pmatrix} 1 & 1 & \cdots & 1 \\ \alpha_1 & \alpha_2 & \cdots & \alpha_m \\ \alpha_1^2 & \alpha_2^2 & \cdots & \alpha_m^2 \\ \vdots & \vdots & \ddots & \vdots \\ \alpha_1^{n-1} & \alpha_2^{n-1} & \cdots & \alpha_m^{n-1} \end{pmatrix}.$$

Hint 5 Use Task 2 and Task 4.

7.4 Exercises

Exercise 7.1 (Zero Norm) Show that $\|\cdot\|_0$ is positive definite and satisfies the triangle inequality, but fails the absolute homogeneity.

Exercise 7.2 (Spark) Let the linear regression data be given as follows:

$$X = \begin{pmatrix} 1 & -1 & 1 & 0 & 0 & 0 \\ 1 & 0 & -1 & 1/2 & 1/2 & 0 \\ 1 & 1 & 0 & -1 & 0 & 0 \\ 1 & 1/3 & 0 & 1/3 & -1 & 1/3 \\ 1 & 1/3 & 1/3 & 1/3 & 0 & -1 \end{pmatrix}, \quad y = \begin{pmatrix} -12 \\ 3 \\ 6 \\ 6 \\ 6 \end{pmatrix}.$$

Show that $w = (0, 12, 0, 6, 0, 0)^T$ solves the optimization problem $(\mathcal{P}_0)$.

Exercise 7.3 (Null Space Property) Given is the matrix

$$X = \begin{pmatrix} 1 & 0 & 1 & 0 \\ 0 & 1 & 1 & 0 \\ 0 & 1 & 0 & 1 \end{pmatrix}.$$

Does X satisfy the null space property of any order?

Exercise 7.4 (Laplace Distribution) Let Z be a random variable following the Laplace distribution $\mathcal{L}(\mu, \tau)$ with the the probability density:

$$p(z) = \frac{1}{2\tau} \cdot e^{-\frac{|z-\mu|}{\tau}},$$

where μ is a location and $\tau > 0$ a scale parameter. Show that the mean of Z equals to μ and its variance to $2\tau^2$.

Exercise 7.5 (Lasso) Consider the following variant of $(\mathcal{L}asso)$:

$$\min_{w} \frac{1}{2} \cdot \|y - X \cdot w\|_2^2 \quad \text{s.t.} \quad \|w\|_1 \leq s, \qquad (\mathcal{V} - \mathcal{L}asso)$$

where $s > 0$ is an upper bound on the Manhattan norm. Show that $(\mathcal{L}asso)$ and $(\mathcal{V} - \mathcal{L}asso)$ are equivalent, i.e. if w solves $(\mathcal{L}asso)$ for a given λ, then it solves $(\mathcal{V} - \mathcal{L}asso)$ for some s, and vice versa.

Exercise 7.6 (ISTA) The following optimization problem is given:

$$\min_{w \in \mathbb{R}^3} \quad \frac{1}{2} \cdot \|y - X \cdot w\|_2^2 + 4 \cdot \|w\|_1,$$

where

$$y = \begin{pmatrix} 1 \\ 4 \end{pmatrix}, \quad X = \begin{pmatrix} 3 & 2 & 6 \\ -2 & 0 & 8 \end{pmatrix}.$$

By computing the spectral norm of X, write the $(\mathcal{ISTA})$ update down. Recover the sparsity pattern of the solution by starting with $w(1) = (1, 1, 1)^T$.

Neural Networks

8

In biology, a *neural network* is a circuit composed of a group of chemically connected or functionally associated neurons. The connections of neurons are modeled by means of weights. A positive weight reflects an excitatory connection, while negative values mean inhibitory connections. All inputs are modified by weights and summed up. This aggregation corresponds to taking linear combinations. Finally, an activation function controls the amplitude of the output. Although each of them being relatively simple, the neurons can build networks with surprisingly high processing power. This gave rise since 1970s to the development of neural networks for solving *artificial intelligence* problems. Remarkable progress in this direction have been achieved particularly in the last decade. As example, we just mention the neural network Leela Chess Zero that managed to win in May 2019 the Top Chess Engine Championship, defeating the conventional chess engine Stockfish in the final. In this chapter, we get to know how the neuron's functioning can be mathematically modeled. This is done on example of classification neurons in terms of the generalized linear regression. The generalization refers here to the use of activation functions. First, we focus on the *sigmoid activation function* and the corresponding *logistic regression*. The training of weights will be based on the minimization of the average cross-entropy arising from the maximum likelihood estimation. We minimize the average cross-entropy by means of the *stochastic gradient descent*. Second, the *threshold activation function* is considered. The corresponding neuron is referred to as *perceptron*. For the latter, the *Rosenblatt learning* is shown to provide a correct linear classifier in finitely many iteration steps. After mentioning the XOR problem, which cannot be handled by perceptrons with one layer, we introduce *multilayer perceptrons*. We point out their importance by stating the *universal approximation theorem*.

© Springer-Verlag GmbH Germany, part of Springer Nature 2021
V. Shikhman, D. Müller, *Mathematical Foundations of Big Data Analytics*,
https://doi.org/10.1007/978-3-662-62521-7_8

8.1   Motivation: Nerve Cells

Neurons are special cells within the nervous system which transmit information to other nerve, muscle, or gland cells. Typically, a neuron has a cell body, an axon, and dendrites, see Fig. 8.1. The cell body contains the nucleus surrounded by the cytoplasm. Dendrites extend from the neuron cell body and receive messages from other neurons. Synapses are the contact points where one neuron communicates with another. The dendrites are covered with synapses formed by the ends of other neurons' axons. When neurons send messages, they transmit electrical impulses along their axons.

Let us present the mathematical model of a neuron, see Fig. 8.2. For that, let the inputs be coming through $m - 1$ dendrites as given by numbers $x_1, \ldots, x_{m-1} \in \mathbb{R}$. Depending on the importance of a particular dendrite, this information is processed within the nucleus:

$$w_0 + x_1 \cdot w_1 + \ldots + x_{m-1} \cdot w_{m-1},$$

where $w_1, \ldots, w_{m-1} \in \mathbb{R}$ are some unknown weights and $w_0 \in \mathbb{R}$ plays the role of a bias. Let us write this linear combination for short as $x^T \cdot w$, where the vectors of inputs and weights are denoted by

$$x = (1, x_1, \ldots, x_{m-1})^T, \quad w = (w_0, w_1, \ldots, w_{m-1})^T.$$

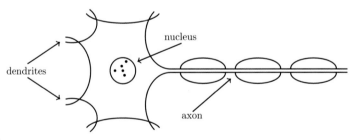

Fig. 8.1 Neuron

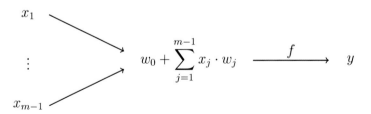

Fig. 8.2 Mathematical model of a neuron

Then, the axon passes on the aggregated signal to other neurons via

$$y = f\left(x^T \cdot w\right),$$

where $y \in \mathbb{R}$ is the output, and $f : \mathbb{R} \to \mathbb{R}$ a properly chosen *activation function*. The latter equation can be interpreted as the *generalized linear regression*.

To become concrete, we want a neuron to be trained for *classification* purposes. For that, let the data set consists, on one hand, of n given input vectors:

$$x_i = \left(1, (x_i)_1, \ldots, (x_i)_{m-1}\right)^T, \quad i = 1, \ldots, n.$$

On the other hand, let them be subdivided into two classes C_{yes} and C_{no}. Equivalently, the data samples are labeled by binary outputs, i.e. for all $i = 1, \ldots, n$ we have:

$$y_i = \begin{cases} 1, \text{ if } i \in C_{yes}, \\ 0, \text{ if } i \in C_{no}. \end{cases}$$

Based on the data set $(x_i, y_i) \in \mathbb{R}^m \times \{0, 1\}$, $i = 1, \ldots, n$, we have to train the weights w. Afterwards, classification can be easily performed by using the *threshold activation function*:

$$f_T(z) = \begin{cases} 1, \text{ if } z \geq 0, \\ 0, \text{ else.} \end{cases}$$

In fact, we just label a newcomer x by setting:

$$y = f_T\left(x^T \cdot w\right).$$

Another possibility would be to approximate the threshold activation function f_T by the *sigmoid activation function*, see Fig. 8.3:

$$f_S(z) = \frac{1}{1 + e^{-z}}.$$

The value $f_S(z)$ can be interpreted as the probability of z to be nonnegative. By using the sigmoid activation function, we can equivalently label a newcomer x as follows:

$$y = \begin{cases} 1, \text{ if } f_S\left(x^T \cdot w\right) \geq 1/2, \\ 0, \text{ else.} \end{cases}$$

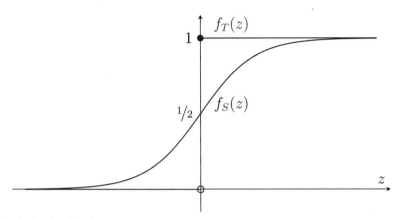

Fig. 8.3 Activation functions

Depending on its activation function, we study how a neuron can successfully learn the weights.

8.2   Results

8.2.1   Logistic Regression

We first focus on the sigmoid activation function. The corresponding neural model is usually referred to as *logistic regression*. After discussing its relation to the logit model, we apply the technique of maximum likelihood estimation which leads us to the minimization of the average cross-entropy. The latter optimization problem will be solved by stochastic gradient descent. This provides us with an algorithm for weights' learning.

8.2.1.1 Logistic Model

In order to justify the use of the sigmoid activation function f_S, let us compute its inverse. For $p \in [0, 1]$ and $z \in \mathbb{R}$ it holds:

$$p = \frac{1}{1 + e^{-z}} \quad \text{if and only if} \quad z = \ln \frac{p}{1 - p}.$$

The latter expression is called the *logit*:

$$\text{logit}(p) = \ln \frac{p}{1 - p}.$$

It takes the logarithm of the odds, which relate the probability p of success to the probability $1 - p$ of failure. Equivalently, we may rewrite the activation of a neuron:

$$y = f_S \left(x^T \cdot w \right)$$

by means of the logit:

$$x^T \cdot w = \text{logit}(y).$$

The logistic regression can be thus said to linearly represent the *log-odds*. It is interesting to note that the logarithm is essential here. Its use is due to the Weber-Fechner law from psychophysics, see e.g. Kandel (2013). The *Weber-Fechner law* relates to human perception, more specifically the relation between the actual change in a physical stimulus and the perceived change. This includes stimuli to all senses, such as vision, hearing, taste, touch, and smell. It states that the subjective sensation is proportional to the logarithm of the stimulus intensity. In our framework, the stimulus intensity is expressed by odds $\frac{y}{1-y}$, and perception $x^T \cdot w$ is assumed to be linear with respect to the unknown weights.

8.2.1.2 Maximum Likelihood Estimation

Let us derive an optimization problem for adjusting weights in the context of logistic regression. This can be done in probabilistic terms by means of *likelihood maximization*. For that, we assume that the labeling process follows the *Bernoulli distribution* $\mathcal{B}(p)$, where the parameter p is given by the sigmoid activation function f_S. This means that the conditional probability of the binary y-label is

$$p(y_i \mid w) = \left(f_S \left(x_i^T \cdot w \right) \right)^{y_i} \cdot \left(1 - f_S \left(x_i^T \cdot w \right) \right)^{1-y_i}, \quad i = 1, \ldots, n.$$

We further assume that the labeling is independent across observations, and eliminate from the notation the dependence on x, to make it look simpler. We emphasize that the data set $x_1, \ldots, x_n$ is considered fixed, so all the randomness associated with y is due to the labeling. The conditional probability density, under the model, of observing the data is their product:

$$p(y \mid w) = \prod_{i=1}^{n} p(y_i \mid w),$$

where, with a slight abuse of notation, we set the vector of labels as

$$y = (y_1, \ldots, y_n)^T.$$

Further, we apply *Bayes theorem* to derive the posterior distribution of the weights w:

$$p(w \mid y) = \frac{p(y \mid w) \cdot p(w)}{p(y)},$$

where $p(y)$ is the probability density of the y-label, and $p(w)$ is a prior distribution of the weights w. Let us assume that all values of w are equally likely, i.e. the prior distribution $p(w)$ is *uniform*. Hence, in order to obtain weights, which better explain the observations, it is reasonable to maximize the so-called *likelihood function*, see e.g. Hendry and Nielsen (2014):

$$L(w) = p(y \mid w).$$

Equivalently, let us consider the *log-likelihood* instead:

$$\ln L(w) = \ln p(y \mid w) = \ln \prod_{i=1}^{n} p(y_i \mid w) = \sum_{i=1}^{n} \ln p(y_i \mid w)$$

$$= \sum_{i=1}^{n} \ln \left(f_S \left(x_i^T \cdot w \right) \right)^{y_i} \cdot \left(1 - f_S \left(x_i^T \cdot w \right) \right)^{1-y_i}$$

$$= \sum_{i=1}^{n} \left(y_i \cdot \ln f_S \left(x_i^T \cdot w \right) + (1 - y_i) \cdot \ln \left(1 - f_S \left(x_i^T \cdot w \right) \right) \right).$$

The *maximum likelihood estimation* for adjusting weights becomes then:

$$\max_{w} \sum_{i=1}^{n} \left(y_i \cdot \ln f_S \left(x_i^T \cdot w \right) + (1 - y_i) \cdot \ln \left(1 - f_S \left(x_i^T \cdot w \right) \right) \right).$$

8.2.1.3 Average Cross-Entropy

First, we rewrite the latter optimization problem as

$$\min_{w} \ H(w) = \frac{1}{n} \cdot \sum_{i=1}^{n} H_i(w), \tag{$\mathcal{ACE}$}$$

where

$$H_i(w) = -y_i \cdot \ln f_S \left(x_i^T \cdot w \right) - (1 - y_i) \cdot \ln \left(1 - f_S \left(x_i^T \cdot w \right) \right).$$

This can be viewed as the cross-entropy from information theory, see e.g. Murphy (2012). Here, the *cross-entropy* measures the dissimilarity between the probability distributions:

$$(y_i, 1 - y_i)^T \quad \text{and} \quad \left(f_S\left(x_i^T \cdot w\right), 1 - f_S\left(x_i^T \cdot w\right) \right)^T.$$

The optimization problem ($\mathcal{ACE}$) can be thus said to minimize the average cross-entropy in the sample. We start to analyze ($\mathcal{ACE}$) by showing that it is a convex optimization problem. The gradient and Hesse matrix of the cross-entropy terms are straightforward to compute, see Exercise 8.2:

$$\nabla H_i(w) = \left(f_S\left(x_i^T \cdot w\right) - y_i \right) \cdot x_i,$$

$$\nabla^2 H_i(w) = f_S\left(x_i^T \cdot w\right) \cdot \left(1 - f_S\left(x_i^T \cdot w\right) \right) \cdot x_i \cdot x_i^T.$$

Since the dyadic products $x_i \cdot x_i^T$ are positive semidefinite, the same holds for the Hesse matrices $\nabla^2 H_i$, $i = 1, \ldots, n$. This does not only provide the convexity of H_i, but also of the average cross-entropy H. This fact allows the application of efficient numerical schemes for solving ($\mathcal{ACE}$).

8.2.1.4 Stochastic Gradient Descent

The technique of stochastic gradient descent is usually applied for minimization of average values, see e.g. Nesterov (2018):

$$H(w) = \frac{1}{n} \cdot \sum_{i=1}^{n} H_i(w).$$

In this framework, the computation of gradients of H is rather challenging. In fact, for the latter we need to average all the gradients of H_i, $i = 1, \ldots, n$, i.e.

$$\nabla H(w) = \frac{1}{n} \cdot \sum_{i=1}^{n} \nabla H_i(w).$$

One step of the usual gradient descent would require the pass through all the data $i = 1, \ldots, n$, which is time consuming. Instead of doing so, a particular index $i \in \{1, \ldots, n\}$ is randomly chosen at each iteration step $t = 1, 2, \ldots$. The *stochastic gradient descent* updates:

$$w(t + 1) = w(t) - \eta \cdot \nabla H_i(w(t)), \tag{$\mathcal{SGD}$}$$

where $\eta > 0$ is an appropriately chosen stepsize. By substituting the previously computed gradient of H_i, we obtain an algorithm for *neural learning*:

$$w(t+1) = w(t) - \eta \cdot \left(f_S \left(x_i^T \cdot w(t) \right) - y_i \right) \cdot x_i,$$

This formula can be interpreted as follows. After an input x_i gets randomly into a neuron, it is aggregated as $x_i^T \cdot w(t)$ by means of the previous weights $w(t)$. The probability of activation $f_S \left(x_i^T \cdot w(t) \right)$ is then compared with the true output y_i. Depending on this comparison, the change of the weights $w(t+1) - w(t)$ is finally set to be proportional to the input x_i.

8.2.1.5 Convergence Analysis of ($\mathcal{SGD}$)

We turn our attention to the convergence analysis of ($\mathcal{SGD}$). For that, let w solve the optimization problem ($\mathcal{ACE}$). We estimate the scaled *Euclidean distance* between $w(t+1)$ and w:

$$\|w(t+1) - w\|_2^2 = \|w(t) - w - \eta \cdot \nabla H_i(w(t))\|_2^2$$

$$= \|w(t) - w\|_2^2 - \eta \cdot 2\nabla^T H_i(w(t)) \cdot (w(t) - w) + \eta^2 \cdot \|\nabla H_i(w(t))\|_2^2.$$

For the last term we derive a uniform upper bound $G > 0$:

$$\|\nabla H_i(w(t))\|_2 = \left\| \left(f_S \left(x_i^T \cdot w(t) \right) - y_i \right) \cdot x_i \right\|_2$$

$$= | \underbrace{f_S \left(x_i^T \cdot w(t) \right)}_{\in [0,1]} - \underbrace{y_i}_{\in \{0,1\}} | \cdot \|x_i\|_2 \leq \|x_i\|_2 \leq \max_{i=1,\ldots,n} \|x_i\|_2 = G.$$

Taking expectation with respect to $i \in \{1, \ldots, n\}$, we thus have:

$$\mathbb{E} \|w(t+1) - w\|_2^2 \leq \mathbb{E} \|w(t) - w\|_2^2 + 2\eta \cdot \mathbb{E} \left(\nabla^T H_i(w(t)) \right) \cdot (w - w(t)) + \eta^2 \cdot G^2.$$

Let us consider the middle term in detail. First, we assume that the choice of i in ($\mathcal{SGD}$) follows the *uniform distribution*. As a consequence, the expectation of the gradients $\nabla^T H_i$ with respect to $i \in \{1, \ldots, n\}$ coincides with their average:

$$\mathbb{E} \left(\nabla H_i(w(t)) \right) = \frac{1}{n} \cdot \sum_{i=1}^{n} \nabla H_i(w(t)),$$

or, equivalently, with the gradient of H itself:

$$\mathbb{E}\left(\nabla H_i(w(t))\right) = \nabla H(w(t)).$$

The convexity of H implies additionally:

$$H(w) \geq H(w(t)) + \nabla^T H(w(t)) \cdot (w - w(t)).$$

Altogether, for the middle term it holds:

$$\mathbb{E}\left(\nabla^T H_i(w(t))\right) \cdot (w - w(t)) = \nabla^T H(w(t)) \cdot (w - w(t)) \leq H(w) - H(w(t)).$$

Substituting above, we get:

$$\mathbb{E}\,\|w(t+1) - w\|_2^2 \leq \mathbb{E}\,\|w(t) - w\|_2^2 + 2\eta \cdot (H(w) - H(w(t))) + \eta^2 \cdot G^2.$$

We rearrange the terms:

$$H(w(t)) - H(w) \leq \frac{1}{\eta} \cdot \frac{\mathbb{E}\,\|w(t) - w\|_2^2 - \mathbb{E}\,\|w(t+1) - w\|_2^2}{2} + \eta \cdot \frac{G^2}{2}.$$

Let us sum up these inequalities over $t = 1, \ldots, T$, and simplify:

$$\sum_{t=1}^{T} H(w(t)) - T \cdot H(w) \leq \frac{1}{\eta} \cdot \frac{\mathbb{E}\,\|w(1) - w\|_2^2 - \mathbb{E}\,\|w(T+1) - w\|_2^2}{2} + \eta \cdot \frac{G^2 \cdot T}{2}.$$

We may skip the first expectation, since the starting point $w(1)$ is arbitrary, but fixed. Moreover, we may estimate the second expectation by zero from below. Additionally, let us divide this inequality by T:

$$\frac{1}{T} \cdot \sum_{t=1}^{T} H(w(t)) - H(w) \leq \frac{1}{\eta} \cdot \frac{\|w(1) - w\|_2^2}{2T} + \eta \cdot \frac{G^2}{2}.$$

By Jensen's inequality, which says that that the convex transformation of an average does not exceed the average taken after the convex transformation, we have:

$$H\left(\bar{w}(T)\right) \leq \frac{1}{T} \cdot \sum_{t=1}^{T} H(w(t)),$$

where the vector of average weights up to iteration T is defined as

$$\bar{w}(T) = \frac{1}{n} \cdot \sum_{t=1}^{T} w(t).$$

Altogether, we arrive at the inequality:

$$H(\bar{w}(T)) - H(w) \leq \frac{1}{\eta} \cdot \frac{\|w(1) - w\|_2^2}{2T} + \eta \cdot \frac{G^2}{2}.$$

In order to adjust the stepsize η, it is convenient to minimize the derived upper bound:

$$\min_{\eta} \frac{1}{\eta} \cdot \frac{\|w(1) - w\|_2^2}{2T} + \eta \cdot \frac{G^2}{2}.$$

A straightforward calculation provides the solution of this optimization problem:

$$\eta = \frac{\|w(1) - w\|_2}{G} \cdot \sqrt{\frac{1}{T}}.$$

By choosing this stepsize, the convergence rate of $(\mathcal{SGD})$ can be given as

$$H(\bar{w}(T)) - H(w) \leq \|w(1) - w\|_2 \cdot G \cdot \sqrt{\frac{1}{T}}.$$

This justifies the use of $(\mathcal{SGD})$ for neural learning. Evaluations of the average cross-entropy $H(\bar{w}(T))$ at the average weights approximate the optimal value $H(w)$ of the optimization problem $(\mathcal{ACE})$. We conclude that the convergence rate is asymptotically vanishing for $T \to \infty$, namely, at the order $1/\sqrt{T}$.

8.2.2 Perceptron

Let us now consider the case of the threshold activation function. The corresponding neural model is known as *perceptron*. By mimicking the stochastic gradient descent, we derive the Rosenblatt learning, which identifies a linear classifier in finite time. Remarkably enough, the Rosenblatt learning fails if the data samples are not linearly separable. This difficulty is overcome by perceptrons with hidden layers.

8.2.2.1 Rosenblatt Learning

Let us assume that the data samples $(x_i, y_i) \in \mathbb{R}^m \times \{0, 1\}$, $i = 1, \ldots, n$, are strictly *linearly separable*. This means that there exists a vector of weights $w \in \mathbb{R}^m$ such that it

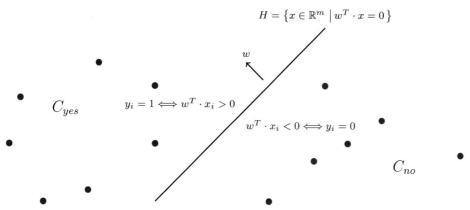

Fig. 8.4 Linear separation

holds for all $i = 1, \ldots, n$:

$$w^T \cdot x_i > 0 \text{ if an only if } y_i = 1, \quad w^T \cdot x_i < 0 \text{ if an only if } y_i = 0.$$

In other words, we can say that the hyperplane

$$H = \left\{ x \in \mathbb{R}^m \,\middle|\, w^T \cdot x = 0 \right\}$$

separates the classes C_{yes} and C_{no} from each other, see Fig. 8.4. How can we learn a *linear classifier* w in a reasonable way? To address this question, we try to mimic the ($\mathcal{SGD}$) update:

$$w(t + 1) = w(t) - \eta \cdot \left(f_S \left(x_i^T \cdot w(t) \right) - y_i \right) \cdot x_i.$$

For that, we exchange the sigmoid activation function f_S by the threshold activation function f_T in ($\mathcal{SGD}$). Moreover, we set the stepsize $\eta = 1$. By doing so, we obtain the *Rosenblatt learning*, also known as the *perceptron algorithm*:

$$w(t + 1) = w(t) - \left(f_T \left(x_i^T \cdot w(t) \right) - y_i \right) \cdot x_i. \tag{$\mathcal{RL}$}$$

The perceptron algorithm was invented by Rosenblatt (1957) at the Cornell Aeronautical Laboratory. The perceptron was intended to be a machine, rather than a program, and while its first implementation was in software, it was subsequently realized as hardware "Mark 1 perceptron". In 1958, the New York Times reported the perceptron to be "the embryo of an electronic computer that [the Navy] expects will be able to walk, talk, see, write, reproduce itself and be conscious of its existence". Aiming to comprehend the importance of the perceptron, let us rewrite the $(\mathcal{RL})$ update:

(1) In case of *correct* classification, i. e. $f_T\left(x_i^T \cdot w(t)\right) = y_i$, the weights in $(\mathcal{RL})$ do not change:

$$w(t + 1) = w(t).$$

(2) In case of *incorrect* classification, i. e. $f_T\left(x_i^T \cdot w(t)\right) \neq y_i$, the weights in $(\mathcal{RL})$ do change as follows:

$$w(t + 1) = \begin{cases} w(t) + x_i, \text{ if } f_T\left(x_i^T \cdot w(t)\right) = 0 \text{ and } y_i = 1, \\ \\ w(t) - x_i, \text{ if } f_T\left(x_i^T \cdot w(t)\right) = 1 \text{ and } y_i = 0. \end{cases}$$

Case (2) can be easily interpreted in terms of neural learning. A misclassification provokes to set the weights' change $w(t+1) - w(t)$ equal to the signed input $\pm x_i$. The sign depends here on whether we mistakenly assigned the input to the class C_{yes} or C_{no}.

8.2.2.2 Convergence Analysis of $(\mathcal{RL})$

Let us show that $(\mathcal{RL})$ stops after finitely many steps providing us with a linear classifier w. For that, we assume that in $(\mathcal{RL})$ the indices $i = 1, \ldots, n$, are changing in *cyclic order*, and we trivially start the iteration process with $w(1) = 0$. Without loss of generality, we may assume for our analysis that all steps $1, \ldots, t$ are of type (2). Otherwise, we just count the updates with incorrect classification. First, we write:

$$\|w(t + 1)\|_2^2 = \|w(t) \pm x_i\|_2^2 = \|w(t)\|_2^2 \pm 2x_i^T \cdot w^T(t) + \|x_i\|_2^2.$$

The last term can be uniformly bounded from above:

$$\|x_i\|_2 \leq \max_{i=1,\ldots,n} \|x_i\|_2 = G.$$

The middle term is always nonpositive due to the misclassification in (2):

$$+x_i^T \cdot w(t) < 0 \text{ if and only if } f_T\left(x_i^T \cdot w(t)\right) = 0,$$

$$-x_i^T \cdot w(t) \leq 0 \text{ if and only if } f_T\left(x_i^T \cdot w(t)\right) = 1.$$

Altogether, we obtain:

$$\|w(t+1)\|_2^2 \leq \|w(t)\|_2^2 + G^2 \leq \ldots \leq \underbrace{\|w(1)\|_2^2}_{=0} + t \cdot G^2 = t \cdot G^2.$$

Further, we have with a linear classifier w:

$$w^T \cdot w(t+1) = w^T \cdot (w(t) \pm x_i) = w^T \cdot w(t) \pm x_i^T \cdot w.$$

The last term here is always positive again due to the misclassification in (2):

$$+x_i^T \cdot w > 0 \text{ if and only if } y_i = 1,$$

$$-x_i^T \cdot w > 0 \text{ if and only if } y_i = 0.$$

Hence, it can be bounded from below by a positive constant:

$$\pm x_i^T \cdot w \geq \min_{i=1,\ldots,n} \pm x_i^T \cdot w = g > 0.$$

Altogether, we obtain:

$$w^T \cdot w(t+1) \geq w^T \cdot w(t) + g \geq \ldots \geq w^T \cdot \underbrace{w(1)}_{=0} + t \cdot g = t \cdot g.$$

The Cauchy-Schwarz inequality, cf. Exercise 2.2, provides on the other hand:

$$t \cdot g \leq w^T \cdot w(t+1) \leq \|w\|_2 \cdot \|w(t+1)\|_2 \leq \|w\|_2 \cdot \sqrt{t \cdot G^2}.$$

This implies that the number of updates (2) in $(\mathcal{RL})$ is finite:

$$t \leq \|w\|_2^2 \cdot \frac{G}{g}.$$

After some time, the iterations in $(\mathcal{RL})$ are exclusively of type (1), i.e. all data samples are correctly classified. We conclude that the perceptron algorithm thus learns a linear classifier.

8.2.2.3 XOR Problem

Although the Rosenblatt learning initially seemed promising, it was quickly shown that perceptrons fail to classify not linearly separable data. As an illustration, we present the meanwhile classical counterexample of XOR. The *XOR problem* models the exclusive disjunction as a logical operation that outputs true only when inputs differ. Let us describe

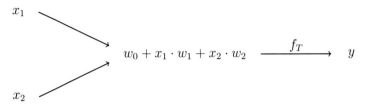

Fig. 8.5 Perceptron with one layer for XOR problem

XOR by the *truth table* which sets out the functional values of logical expressions on each
of their functional arguments:

input x	$(1, 1)$	$(1, 0)$	$(0, 1)$	$(0, 0)$
output y	0	1	1	0

We show that a perceptron with two inputs x_1 and x_2 cannot solve the XOR problem. In
other words, there do not exist weights w_0, w_1, and w_2 such that the neuron activates if and
only if x_1 and x_2 differ, see Fig. 8.5. If this were possible, we would have by substituting
each of four data samples:

$$f_T (w_0 + 0 \cdot w_1 + 0 \cdot w_2) = 0, \quad f_T (w_0 + 1 \cdot w_1 + 0 \cdot w_2) = 1,$$

$$f_T (w_0 + 0 \cdot w_1 + 1 \cdot w_2) = 1, \quad f_T (w_0 + 1 \cdot w_1 + 1 \cdot w_2) = 0.$$

Equivalently, we obtain:

$$w_0 < 0, \quad w_0 + w_1 \geq 0, \quad w_0 + w_2 \geq 0, \quad w_0 + w_1 + w_2 < 0.$$

From the two first inequalities it follows that $w_1 > 0$, whereas the two other inequalities
provide $w_1 < 0$, a contradiction. The impossibility for a perceptron to solve the XOR
problem comes from the fact that the corresponding classes C_{yes} and C_{no} are not linearly
separable, see Fig. 8.6. This caused the research of neural networks to stagnate for decades,
before it was recognized that perceptrons with multiple layers have greater processing
power. In particular, it is not hard to construct a perceptron with two layers that solves the
XOR problem, see Fig. 8.7. We refer to Exercise 8.6 for further details.

8.2.2.4 Multilayer Perceptron

In general, a *multilayer perceptron* consists of at least three layers of nodes: an input layer,
a hidden layer and an output layer, see Fig. 8.8. The inputs $x_1, \ldots, x_{m-1} \in \mathbb{R}$ are linearly
processed by hidden neurons $\ell = 1, \ldots, k$:

$$w_0^\ell + x_1 \cdot w_1^\ell + \ldots + x_{m-1} \cdot w_{m-1}^\ell,$$

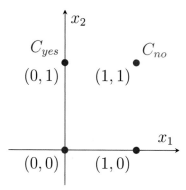

Fig. 8.6 Failure of linear separation in XOR problem

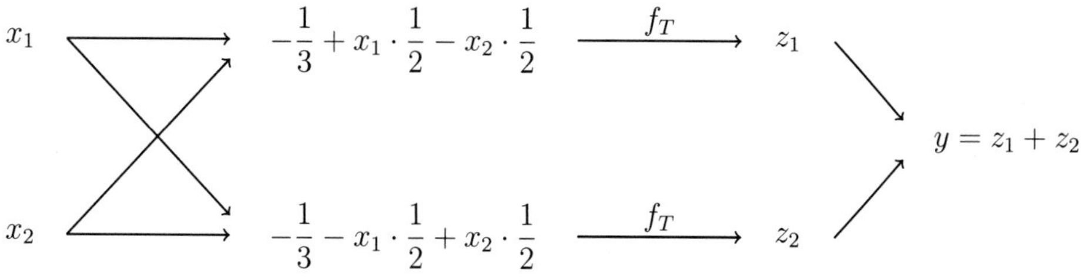

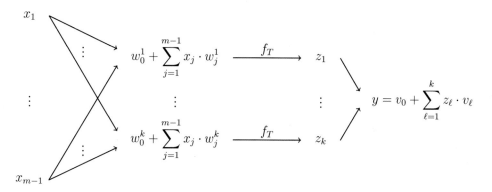

Fig. 8.7 Perceptron with two layers for XOR problem

Fig. 8.8 Multilayer perceptron

where the weights $w_0^\ell, w_1^\ell, \ldots, w_{m-1}^\ell \in \mathbb{R}$ correspond to the ℓ-th hidden neuron. As above, we write this linear combination for short as $x^T \cdot w^\ell$ in vector form, where

$$x = (1, x_1, \ldots, x_{m-1})^T, \quad w^\ell = \left(w_0^\ell, w_1^\ell, \ldots, w_{m-1}^\ell\right)^T.$$

Each of the hidden neurons passes on the aggregated signal, i. e.

$$z_\ell = f_T\left(x^T \cdot w^\ell\right),$$

where $z_\ell \in \mathbb{R}$ is the output of the ℓ-th hidden neuron. Finally, the output $y \in \mathbb{R}$ of the multilayer perceptron is given by the linear combination:

$$y = v_0 + z_1 \cdot v_1 + \ldots + z_k \cdot v_k,$$

where $v_1, \ldots, v_k \in \mathbb{R}$ weight the importance of hidden neurons and $v_0 \in \mathbb{R}$ plays the role of a bias. We also write the output $y = z^T \cdot v$ as scalar product, where

$$z = (1, z_1, \ldots, z_k)^T, \quad v = (v_0, v_1, \ldots, v_k)^T.$$

Overall, we obtain the following generalized linear regression:

$$y = \sum_{\ell=0}^{k} f_T\left(x^T \cdot w^\ell\right) \cdot v_\ell.$$

Note that here not only the weights $w^1, \ldots, w^k \in \mathbb{R}^m$, but also $v_1, \ldots, v_k \in \mathbb{R}$ are to be properly adjusted. This flexibility of weights' adjustment makes multilayer perceptron a rather powerful tool for modeling purposes. What kind of models can be resolved by multilayer perceptrons? It turns out that multilayer perceptrons are able to arbitrarily well approximate any continuous model. This is due to the *universal approximation theorem* for neural networks. More precisely, let a general model be given as

$$y = F(x),$$

where $F : K \to \mathbb{R}$ is a continuous function and $K \subset \mathbb{R}^m$ is a compact subset. Then, for every precision $\varepsilon > 0$ there exist weights $w^1, \ldots, w^k \in \mathbb{R}^m$ and $v_1, \ldots, v_k \in \mathbb{R}$, such that it holds:

$$\left| F(x) - \sum_{\ell=0}^{k} f_T\left(x^T \cdot w^\ell\right) \cdot v_\ell \right| \le \varepsilon \quad \text{for all } x \in K.$$

One of the first versions of this theorem was proved by Cybenko (1989) for sigmoid activation functions at the hidden layer. Later on it has been recognized that already threshold activation functions suffice. Hornik (1991) showed that it is not the specific choice of the activation function, but rather the multilayer architecture itself which gives neural networks the potential of being universal approximators. Since any continuous model can be approximated by a multilayer perceptron, the use of neural networks became rather popular since then. However, it is worth to mention that the number k of hidden neurons in the universal approximation theorem depends on the particular problem, and may be huge. This motivates the introduction of neural networks with more sophisticated structure, such as convolutional, recursive, recurrent, and long short-term memory neural networks to name the few, see e.g. Haykin (2011) for more details. Another obstacle is that the learning of weights for a neural network is normally based on the solution of a highly nonconvex optimization problem. This hampers the derivation of convergence rates aiming to at least theoretically explain why neural networks do perform so well in practice. We conclude that neural networks remain an area of very active research till nowadays.

8.3   Case Study: Spam Filtering

We consider a system for *spam filtering*. Its main task is to classify incoming emails as valid or spam. Let the binary variable $y \in \{0, 1\}$ denote the true class of an incoming email, i. e. $y = 0$ means that the email is valid, and $y = 1$ means that the email is spam. Let us further represent emails within the *bag-of-words model*. This is done by disregarding grammar and even word order, but keeping the words' multiplicity. An email is represented as a vector $x \in \{0, 1\}^m$, where m is the number of words in the dictionary. For its entry it holds $x_j = 1$ if the j-th word is contained in the email, and $x_j = 0$ otherwise. Aiming to predict whether an email x is valid or spam, we use a linear filter $w \in \mathbb{R}^m$ by labeling:

$$y = \begin{cases} 1, & \text{if } x^T \cdot w \geq 0, \\ 0, & \text{else.} \end{cases}$$

In order to guarantee the implementability, we additionally assume that spam filters are bounded with respect to the Euclidean norm. For short, we write then $w \in B_R$, where the ball of radius $R > 0$ is denoted by

$$B_R = \left\{ w \in \mathbb{R}^m \mid \|w\|_2 \leq R \right\}.$$

If a data set given, it is straightforward to construct a spam filtering system based on the logistic regression, and to subsequently train it, say, by using the $(\mathcal{SGD})$ algorithm. However, a spam filtering system has also to cope with adversarial data generation and to be dynamically adjusted with the varying input. Thus, it is reasonable to apply the *online learning* framework from Chap. 2 instead.

Task 1 By using the cross-entropy, derive the quality measure for a spam filter w in presence of an email x and its true label y. Assume that prediction probability is modeled by means of the sigmoid activation function f_S.

Hint 1 The cross-entropy

$$H(w, x, y) = -y \cdot \ln f_S\left(x^T \cdot w\right) - (1 - y) \cdot \ln\left(1 - f_S\left(x^T \cdot w\right)\right)$$

measures the dissimilarity between the two distributions:

$$(y, 1 - y)^T \quad \text{and} \quad \left(f_S\left(x^T \cdot w\right), 1 - f_S\left(x^T \cdot w\right)\right)^T.$$

Task 2 Derive and interpret the formula of the average regret by using the cross-entropy as a quality measure for spam filters.

Hint 2 The average regret is

$$\mathcal{R}(T) = \frac{1}{T} \cdot \sum_{t=1}^{T} f_t(w(t)) - \min_{w \in B_R} \frac{1}{T} \cdot \sum_{t=1}^{T} f_t(w),$$

where we have

$$f_t(w) = H(w, x(t), y(t)).$$

Note that $x(t)$ represents the t-th email, and $y(t)$ denotes its true label.

Task 3 Show that the loss functions $f_t(w) = H(w, x(t), y(t))$ have uniformly bounded gradients with respect to the Euclidean norm.

Hint 3 It holds:

$$\nabla f_t(w) = \left(f_S\left(x^T(t) \cdot w\right) - y(t)\right) \cdot x(t).$$

For all $w \in B_R$ and $t = 1, 2, \ldots$ it easily follows:

$$\|\nabla f_t(w)\|_2 \le \sqrt{m} = G.$$

Task 4 Determine the diameter of B_R with respect to the *Euclidean prox-function*:

$$D = \sqrt{\max_{u, w \in B_R} \frac{1}{2}\|u\|_2^2 - \frac{1}{2}\|w\|_2^2}.$$

Hint 4 It holds:

$$D = \frac{R}{\sqrt{2}}.$$

Task 5 Derive the explicit formula for the *Euclidean projection* of a vector u on B_R:

$$\text{proj}_{B_R}(u) = \arg\min_{w \in B_R} \|u - w\|_2.$$

Hint 5 It holds:

$$\text{proj}_{B_R}(u) = \begin{cases} u, & \text{if } \|u\|_2 \leq R, \\ R \cdot \dfrac{u}{\|u\|_2}, & \text{if } \|u\|_2 > R. \end{cases}$$

Task 6 Apply the *online gradient descent* for spam filtering, see Exercise 2.7:

$$w(t+1) = \text{proj}_{B_R}(w(t) - \eta \cdot \nabla f_t(w(t))), \quad w(1) = \text{proj}_{B_R}(0).$$

Derive the convergence rate for the average regret. How should the stepsize η be adjusted?

Hint 6 ($\mathcal{OGD}$) for spam filtering reads:

$$w(t+1) = \text{proj}_{B_R}\left(w(t) - \eta \cdot \left(f_S\left(x^T(t) \cdot w(t)\right) - y(t)\right) \cdot x(t)\right), \quad w(1) = 0.$$

The corresponding convergence rate and the stepsize are

$$\mathcal{R}(T) \leq R \cdot \sqrt{\frac{m}{T}}, \quad \eta = R \cdot \sqrt{\frac{1}{m \cdot T}}.$$

8.4 Exercises

Exercise 8.1 (Neural Network) A neural network is designed for marketing purposes. As inputs, it assigns to customers their age x_1, income x_2, and number of previous purchases x_3. The output $y \in \{0, 1\}$ indicates whether the advertisement leads to a purchase. The weights of the inputs have already been estimated on the basis of customer data:

$$w_1 = -0.1, \quad w_2 = 0.6, \quad w_3 = 0.7.$$

The bias is trivial with $w_0 = 0$. The neural network has one layer and is activated by the sigmoid function. What is the probability that the following customers will respond to the advertisement with a purchase?

	Customer 1	Customer 2	Customer 3
Age x_1	20	30	40
Income x_2	6	5	1
Number of previous purchases x_3	1	0	3

Exercise 8.2 (Sigmoid Activation Function) Given is the following initial value problem:

$$f' = f \cdot (1 - f), \quad f(0) = \frac{1}{2}.$$

Show that it admits the unique solution, namely, the sigmoid activation function.

Exercise 8.3 (Logistic Distribution) Show that the sigmoid activation function is a cumulative distribution function:

$$f_S(z) = \mathbb{P}(\varepsilon \leq z),$$

where the random variable ε follows the *logistic distribution* $\mathcal{L}(0, 1)$ with the zero location and scale parameter equal to one. The probability density of the logistic distribution is

$$p(z) = \frac{e^{-z}}{\left(1 + e^{-z}\right)^2}.$$

Show that ε has zero mean, and its variance equals to $\pi^2/3$.

Exercise 8.4 (Latent-Variable Model) Let us consider the following linear regression model which links a latent endogenous variable $y^* \in \mathbb{R}$ with exogenous variables $x \in \mathbb{R}^m$:

$$y^* = x^T \cdot w + \varepsilon,$$

where $w \in \mathbb{R}^m$ is the vector of unknown weights, and the random error ε follows the logistic distribution $\mathcal{L}(0, 1)$. The classification of a newcomer x is given by labeling:

$$y = \begin{cases} 1, & \text{if } y^* \geq 0, \\ 0, & \text{else.} \end{cases}$$

Show that for the probability of labeling x by $y = 1$ it holds:

$$\mathbb{P}(y = 1 \,|\, x) = f_S(x^T \cdot w).$$

Exercise 8.5 (Cross-Entropy) Compute the gradient and Hesse matrix of the cross-entropy for all $i = 1, \ldots n$:

$$H_i(w) = -y_i \cdot \ln f_S\left(x_i^T \cdot w\right) - (1 - y_i) \cdot \ln\left(1 - f_S\left(x_i^T \cdot w\right)\right).$$

Exercise 8.6 (Multilayer Perceptron) Show that the multilyer perceptron from Fig. 8.7 solves the XOR problem.

Decision Trees

9

Decision tree learning is one of the predictive modelling approaches widely used in the fields of data mining and machine learning. It uses a *decision tree* to go through testing an object in the nodes to conclusions about its target variable's value in the leaves. Decision trees are among the most popular machine learning algorithms given their intelligibility and simplicity. The applications range from the prediction of Titanic survival to the artificial intelligence chess playing. In this chapter, we focus on the classification decision trees, so that their leaves represent class labels. The quality of such decision trees is measured by means of both the *misclassification rate* on the given data, and the *average external path length*. Especially for identification decision trees with zero misclassification rate, we show that finding those with the minimal average external path length is an *NP-complete problem*. Due to this negative theoretical result, *top-down and bottom-up heuristics* are proposed to nevertheless construct decision trees whose quality is sufficient at least from the practical point of view. Based on various *generalization errors*, such as train error, entropy, and Gini index, we present the *iterative dichotomizer* algorithm for this purpose. The iterative dichotomizer splits at each step the data set by maximizing the gain derived from the chosen generalization error. Afterwards, we briefly elaborate on the *pruning* of decision trees.

9.1 Motivation: Titanic Survival

The RMS Titanic sank in the early morning hours of 15 April 1912 in the North Atlantic Ocean, four days into the ship's maiden voyage from Southampton to New York. The largest ocean liner in service at the time, Titanic had an estimated 2224 people on board when it struck an iceberg. Unfortunately, there were not enough lifeboats for everyone available, resulting in the death of 1502 passengers and crew. The sinking of Titanic 2 h and

© Springer-Verlag GmbH Germany, part of Springer Nature 2021
V. Shikhman, D. Müller, *Mathematical Foundations of Big Data Analytics*,
https://doi.org/10.1007/978-3-662-62521-7_9

40 min later marks one of the deadliest peacetime marine disasters in history. Since then the Titanic catastrophe has never faded from the world's imagination. Why does the Titanic continue to be what flashes into people's minds whenever the word "disaster" comes up? It is virtually the only disaster that is perpetually remembered, commemorated, and even celebrated. The answer has to do with the drama of choice, not with the brute facts of the disaster itself. As CNN covers the story by commemorating the Titanic's centenary in 2012: "Its cast of characters included people of every rank and station and personality. The cast was large enough to represent the human race, yet small enough to form a self-contained society, in which individuals could see what other individuals were doing, and think carefully about their own responses. The Titanic had what every great drama needs: a relentless focus on the supreme choices of individual lives".

This motivates to predict the survival of passengers aboard the RMS Titanic just by using information on passenger's gender, age, family, class etc. As prediction model, we choose decision trees. A decision tree is a flowchart-like structure in which each node represents a test on an attribute, each branch represents the outcome of the test, and each leaf represents a class label, see Fig. 9.1. The nodes split the passengers according to the corresponding features, starting by gender (male or female) and following by age (older or younger than 9.5 years), and siblings (more or less than 3). The leafs of this decision tree display the prediction for "died" or "survived". The survival probability conditioned on the branch is given on the left, and the percentage of observations on the right of an every leaf. E.g., from all males older than 9.5 years, which made up 60% of the passengers, nearly 17% survived. From all males younger than 9.5 years having less than 3 siblings

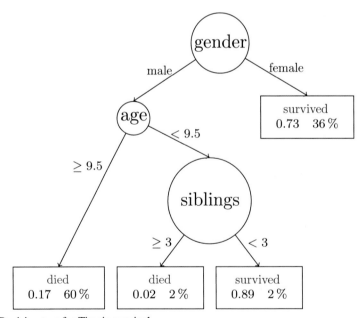

Fig. 9.1 Decision tree for Titanic survival

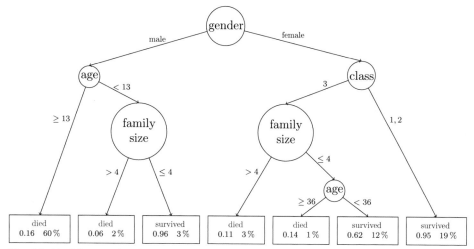

Fig. 9.2 Decision tree for Titanic survival

aboard, which made up 2% of the passengers, around 89% survived. By summarizing, the chances of survival were good if you were a female or a male younger than 9.5 years with strictly less than 3 siblings. From this example we see the main advantage of using decision trees for classification purposes: they are easy to understand and allow a straightforward interpretation. However, depending on the features chosen for data splitting, decision trees for one and the same problem may considerably differ, see Fig. 9.2. Here, we additionally examined the survival chances of females conditioned on the occupied cabin class. It turns out that most of them from the third class had fewer chances to survive, in particular, if having large families of more than 4 members or being older than 36 years. The decision tree on Fig. 9.2 reveals the circumstances of the Titanic's rescue operation. The ship only carried enough lifeboats for slightly more than half of the people on board. In this respect, the most significant aspect of the rescue effort was the "women and children first" policy followed in the majority of lifeboat loadings. Additionally, first and second class passengers could reach the lifeboats much faster than those from the third class. This was due to the fact that first and second class cabins and facilities were closer to the upper decks, where the lifeboats were stored, than to the lower decks. It should thus not come as a surprise that survival probability was heavily skewed towards woman, children and in general those of the first and second classes.

Based on the Titanic example, we present the formal definition of a decision tree for classification purposes. For that, let an abstract set of objects $\mathcal{O}$ be under consideration. We assume that binary tests can be performed on the objects. These tests will be represented by means of mappings $T : \mathcal{O} \rightarrow \{0, 1\}$. We say that an object $x \in \mathcal{O}$ fails the test if $T(x) = 0$, and passes it if $T(x) = 1$. In other words, a test signals if an object is particularly featured or not. A *decision tree* D is a predictor that associates one of the classes $C_1, \ldots, C_\ell$ with an object x by traveling from the root of a tree to one of its leafs L. At the root and all internal nodes a test T is specified, and the leafs L classify the object. The classification procedure is as follows. First, the test specified at the root is applied to a newcomer x. If it is false, i.e. $T(x) = 0$, one takes the left branch, otherwise the right, i.e. $T(x) = 1$. This procedure is repeated at the root of each successive subtree until a leaf L is reached. The latter defines the class $D(x) \in \{C_1, \ldots, C_\ell\}$ of the newcomer x. For simplicity, we assume that there are finitely many tests:

$$\mathcal{T} = \{T_1, \ldots, T_m\}.$$

In order to design a reliable decision tree, a data set of already classified objects is given:

$$\mathcal{X} = \{x_1, \ldots, x_n\} \subset \mathcal{O}.$$

This means that for any object $x_i \in \mathcal{X}$ there exists the unique class C_{k_i} which x_i belongs to, i.e. $x_i \in C_{k_i}$ with $k_i \in \{1, \ldots, \ell\}$. Our aim is to learn a decision tree D that, on the one hand, fits the data well enough and, on the other hand, is not too large. Both objectives need to be quantified:

- The *misclassification rate* measures the quality of the decision tree D:

$$\mu(D) = \frac{\#\{i \in \{1, \ldots, n\} \mid D(x_i) \neq C_{k_i}\}}{n}.$$

  The misclassification rate is thus the percentage of those data samples within the training set $\mathcal{X}$, which are incorrectly classified by the decision tree D.
- The *average external path length* captures the size of the decision tree D:

$$\rho(D) = \frac{1}{n} \cdot \sum_{i=1}^{n} \rho(x_i),$$

  where $\rho(x_i)$ measures the path length from the root to the leaf classifying x_i. The average external path length thus counts the expected number of tests needed for an object from the training set $\mathcal{X}$ to be classified by the decision tree D.

Overall, given a data set $\mathcal{X}$, we are searching for a decision tree D where both the misclassification rate $\mu(D)$ and the average external path length $\rho(D)$ are minimal. Unfortunately, this two objectives are coupled. Decision trees of relatively small size normally cause high misclassification rates, and the minimization of the misclassification rate produces decision trees with long external paths. In what follows, we study the complexity of finding optimal decision trees, as well as the heuristics towards this goal.

9.2   Results

9.2.1   NP-Completeness

Hyafil and Rivest (2009) demonstrated that constructing optimal binary decision trees is an NP-complete problem, where optimality refers to the the minimal expected number of tests required to identify the unknown object. Let us briefly present the corresponding ideas.

9.2.1.1 Decision Tree Problem

First, we note that *identification* is a special case of classification. To see this, we set $\mathcal{O} = \mathcal{X}$, and let the number of classes coincide with the number of objects, i. e. $\ell = n$. Moreover, let the classes contain exactly one object each. Without loss of generality, we may set:

$$C_i = \{x_i\}, \quad i = 1, \ldots, n.$$

We thus have that $k_i = i$ for all $i = 1, \ldots, n$. Further, let us consider decision trees with zero misclassification rate, i.e. $\mu(D) = 0$. This means that all objects are correctly identified:

$$D(x_i) = C_i, \quad i = 1, \ldots, n.$$

Consequently, there are n leafs which uniquely identify the objects. The formula for the average external path length then reads:

$$\rho(D) = \frac{1}{n} \cdot \sum_{i=1}^{n} \rho(x_i),$$

where $\rho(x_i)$ measures the path length from the root to the leaf naming x_i. The *decision tree problem* $\mathrm{DT}(\mathcal{X}, \mathcal{T}, \rho)$ is to determine—in case of existence—an identification decision tree D with the average external path length $\rho(D)$ less than or equal to $\rho > 0$, given the data set $\mathcal{X}$ and the set of binary tests $\mathcal{T}$. We want to show that the decision tree problem is hard to solve in a certain sense. For that, we need to delve into the foundations of complexity theory.

9.2.1.2 P versus NP

The *P versus NP* problem is a major unsolved problem in computer science. Loosely speaking, it asks whether every problem whose solution can be quickly verified can also be solved quickly. It is one of the seven Millennium Prize Problems selected by the Clay Mathematics Institute, each of which carries a prize of one million USD for the first correct solution. The informal term quickly, used above, means the existence of an algorithm solving the task that runs in polynomial time. In other words, the time to complete the task varies as a polynomial function on the size of the input to the algorithm. In our case of the decision tree problem $DT(\mathcal{X}, \mathcal{T}, \rho)$ the input would comprise of the data set $\mathcal{X}$, the test set $\mathcal{T}$, and the upper bound ρ on the expected number of tests required for the objects' identification. The general class of questions for which some algorithm can provide an answer in *polynomial time* is called P. For some questions, there is no known way to find an answer quickly, but if one is provided with an answer, it is possible to verify it quickly. The class of questions for which an answer can be verified in polynomial time is called NP. The latter stands for *nondeterministic polynomial time*.

Let us examine whether the decision tree problem $DT(\mathcal{X}, \mathcal{T}, \rho)$ falls into the complexity class NP. Is it possible to verify in polynomial time that an identification decision tree D solves $DT(\mathcal{X}, \mathcal{T}, \rho)$? For that, we try to successively identify the objects $x_1, \dots, x_n$ by means of the decision tree D. If for x_i we arrive at the leaf naming x_j with $i \neq j$, then the candidate D for the solution of $DT(\mathcal{X}, \mathcal{T}, \rho)$ has to be rejected. During the identification process we count the total number M of tests applied. If at some moment it happens that $\frac{M}{n} > \rho$, then D has also to be rejected. It remains to note that either after at most $\lfloor \rho \cdot n \rfloor + 1$ tests we verified that D does not solve the decision problem $DT(\mathcal{X}, \mathcal{T}, \rho)$, or after having successfully identified all n objects from $\mathcal{X}$ we convince ourselves in the opposite. In both cases the verification can be performed in polynomial time.

9.2.1.3 Exact Cover Problem

To attack the P versus NP question, the concept of NP-completeness is very useful. An *NP-complete problem* is a particular NP problem to which any other NP problem can be reduced in polynomial time. A consequence of this definition is that if we had a polynomial time algorithm for an NP-complete problem, we could solve all NP problems in polynomial time, i.e. showing that P $=$ NP. Informally, an NP-complete problem is at least as tough as any other problem in NP. Nowadays, most of the researchers believe that P $\neq$ NP. A key reason for this belief is that after decades of studying these problems no one has been able to find a polynomial-time algorithm for any of more than 3000 important NP-complete problems. It is also intuitively argued that the existence of problems, that are hard to solve, but for which the solutions are easy to verify, matches real-world experience.

To get better acquainted with NP-complete problems, let us present a typical one, called exact cover. For that, let $\mathcal{Y} = \{y_1, \dots, y_n\}$ be a finite set, and $\mathcal{E} = \{E_1, \dots, E_m\}$ be a family of subsets of $\mathcal{Y}$, each of them with three elements, i.e.

$$E_j \subset \mathcal{Y} \quad \text{and} \quad |E_j| = 3, \quad j = 1, \dots, m.$$

An *exact cover* is a subcollection $\mathcal{E}^* \subset \mathcal{E}$, such that each element $y \in \mathcal{Y}$ is contained in exactly one of its subsets $E \in \mathcal{E}^*$. One says that each element in $\mathcal{Y}$ is covered by exactly one subset from $\mathcal{E}^*$. The *exact cover problem* $EC3(\mathcal{Y}, \mathcal{E})$ is to find an exact cover for a set $\mathcal{Y}$ by means of the given family $\mathcal{E}$ of subsets or to recognize that such exact cover does not exist. The exact cover—with no restriction on the size of E_j's—is one of the famous 21 NP-complete problems provided by Karp (1972), see also Exercise 9.3.

9.2.1.4 Polynomial-Time Reduction

Now, we are ready to prove that the decision tree problem is NP-complete. For that, we show that another NP-complete problem, namely the exact cover, can be *reduced* to it *in polynomial time*. That is, given an exact cover instance $EC3(\mathcal{Y}, \mathcal{E})$, we provide a polynomial-time algorithm aiming to construct a corresponding decision tree instance $DT(\mathcal{X}, \mathcal{T}, \rho)$, so that solving the latter will provide us with an exact cover. The reduction of $EC3(\mathcal{Y}, \mathcal{E})$ to $DT(\mathcal{X}, \mathcal{T}, \rho)$ is as follows:

(i) The set of objects is defined as

$$\mathcal{X} = \mathcal{Y} \cup \{a, b, c\},$$

where $a, b, c \notin \mathcal{Y}$ are three additional objects.

(ii) Given a subset $E \subset \mathcal{X}$, the incidence test for $x \in \mathcal{X}$ is defined as

$$T_E(x) = \begin{cases} 0, & \text{if } x \notin E, \\ 1, & \text{if } x \in E. \end{cases}$$

We consider the incidence tests which correspond to the subsets from the collection $\mathcal{E}$ and to the singleton subsets of $\mathcal{X}$:

$$\mathcal{T} = \{T_E \mid E \in \mathcal{E} \text{ or } |E| = 1\}.$$

(iii) The upper bound ρ is taken as the minimal average external path length over all identification decision trees D with the data set $\mathcal{X}$ and the set of incidence tests $\mathcal{T}$, i.e.

$$\rho = \min \{\rho(D) \mid D \text{ identifies } \mathcal{X} \text{ by means of } \mathcal{T}\}.$$

Let us examine whether the above reduction can be performed in polynomial time with respect to the input $\mathcal{Y}$ and $\mathcal{E}$ of the exact cover problem. In (i) we just added 3 objects a, b, c, and, in (ii) the number of tests grew by $n + 3$ singleton subsets of $\mathcal{X}$. The polynomial-time computation of ρ in (iii) is more involved. For that, we consider the decision tree $\bar{D}$ which identifies all $n + 3$ objects from $\mathcal{X}$ by singleton tests, see Fig. 9.3. For its average external path length we obtain:

$$\rho\left(\bar{D}\right) = \frac{1 + \ldots + (n+2) + (n+2)}{n+3} = \frac{\frac{(n+3)\cdot(n+2)}{2} + (n+2)}{n+3} = \frac{\frac{(n+5)\cdot(n+2)}{2}}{n+3}.$$

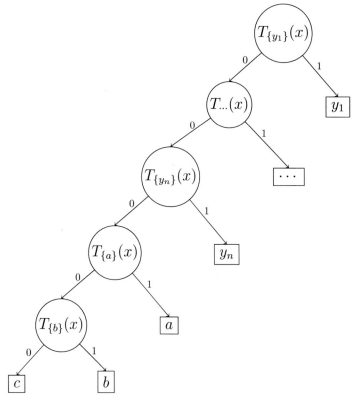

Fig. 9.3 Decision tree $\bar{D}$

Now, we successively solve the decision tree problems DT$(\mathcal{X}, \mathcal{T}, \rho_s)$, where

$$\rho_s = \frac{\frac{(n+5)\cdot(n+2)}{2} - s}{n+3}, \quad s = 0, 1, \ldots.$$

For any s the decision tree problem DT$(\mathcal{X}, \mathcal{T}, \rho_s)$ provides—in case of existence—an identification decision tree with the average external path length less than or equal to ρ_s. In view of $\rho_0 = \rho(\bar{D})$, the decision tree problem DT$(\mathcal{X}, \mathcal{T}, \rho_0)$ is feasible, and, hence, solvable. Assume that for some s the decision tree problem DT$(\mathcal{X}, \mathcal{T}, \rho_s)$ remains solvable, but DT$(\mathcal{X}, \mathcal{T}, \rho_{s+1})$ does not. This would give us the minimal average external path length:

$$\rho = \rho_s.$$

Note that for computing of ρ we need to solve not more than $\frac{(n+5)\cdot(n+2)}{2}$ instances of the decision tree problem. This number of calls is polynomial in n.

9.2.1.5 Minimal External Path Length

On the way of showing how the solution of $DT(\mathcal{X}, \mathcal{T}, \rho)$ embodies the solution of $EC3(\mathcal{Y}, \mathcal{E})$, we study decision trees with minimal *external path lengths*. For that, let $f(n)$ denote the minimal external path length of an identification decision tree on an n-element set, where all 1- and 3-element subsets are available as tests. We give the first few values of f, see Exercise 9.2:

n	1	2	3	4	5	6	7	8
$f(n)$	0	2	5	9	12	16	21	25

Let us derive a recursive representation for f. For that, we test a newcomer by an i-element subset with $i = 1$ or 3. Then, the left subtree requires $f(n - i)$ further tests, and $f(i)$ additional tests come from the right subtree. Besides, we have already performed one test for n leafs, so that we sum up together:

$$f(n) = \min_{i=1,3} \ f(n - i) + f(i) + n.$$

Let us show by induction that for all $n \geq 4$ it holds:

$$f(n) - f(n - 1) \geq 3.$$

- The *base of induction* is clear, since we have:

$$f(4) - f(3) = 9 - 5 = 4 > 3, \quad f(5) - f(4) = 12 - 9 = 3.$$

- For *induction hypothesis* we assume that the inequality holds up to $n - 1$ with $n \geq 7$.
- It remains to accomplish the *induction step*. We apply the induction hypothesis twice to obtain for $k \geq 6$:

$$\underbrace{f(k - 1)}_{\geq f(k-2)+3} + \underbrace{f(1)}_{=1} + k \geq \underbrace{f(k - 2)}_{\geq f(k-3)+3} + 4 + k \geq f(k - 3) + 7 + k > f(k - 3) + \underbrace{f(3)}_{=5} + k.$$

This means that in the recursive representation of f the minimum is taken at $i = 3$, i.e.

$$f(k) = f(k - 3) + f(3) + k.$$

By using the latter and the induction hypothesis once more, we finally get for $n \geq 7$:

$$f(n) - f(n-1) = (f(n-3) + f(3) + n) - (f(n-4) + f(3) + n - 1)$$

$$= \underbrace{f(n-3) - f(n-4)}_{\geq 3} + 1 > 3,$$

and we are done with the proof by induction.

Now, it is easy to see that the root of the optimal decision tree always selects a 3-element subset as a test, at least for $n \geq 6$. In fact, we have:

$$\underbrace{f(n-1)}_{\geq f(n-2)+3} + \underbrace{f(1)}_{=1} + n \geq \underbrace{f(n-2)}_{\geq f(n-3)+3} + 4 + n \geq f(n-3) + 7 + n > f(n-3) + \underbrace{f(3)}_{=5} + n,$$

and in the recursive representation of f the minimum is achieved at $i = 3$.

9.2.1.6 Optimal Decision Tree

Let D be an identification decision tree solving the problem $\mathrm{DT}(\mathcal{X}, \mathcal{T}, \rho)$ as given by (i)–(iii). We want to show that it has the structure from Fig. 9.4, where the subcollection

$$\mathcal{E}^* = \left\{ E_{j_1}, \ldots, E_{j_r} \right\}$$

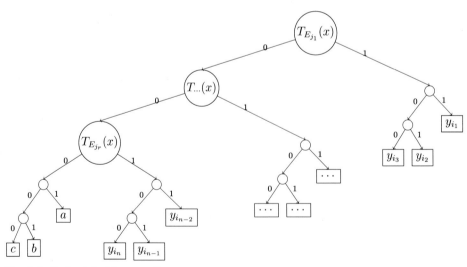

Fig. 9.4 Optimal decision tree D

forms an exact cover of $\mathcal{Y}$ and the singleton tests are used to distinguish objects within each triple. Due to (iii), D minimizes the average external path length, or equivalently, the external path length itself. As we have shown previously, identification decision trees with minimal external path lengths always select at the root 3-element subsets as tests. By applying this argument to all subtrees of D with at least six objects—this is why we added three dummy objects a, b, c,—we obtain its form as given in Fig. 9.4. Furthermore, any optimal decision tree must embody the solution to the corresponding exact cover problem, because only in this way can the root of each sufficiently large subtree be a test on a 3-element subset of $\mathcal{X}$. This provides that solving DT$(\mathcal{X}, \mathcal{T}, \rho)$ gives us a solution of EC3$(\mathcal{Y}, \mathcal{E})$. Hence, the NP-complete exact cover problem is reducible to the decision tree problem in polynomial time, implying that the later is NP-complete too.

9.2.2 Top-Down and Bottom-Up Heuristics

Since the construction of optimal decision trees turns out to be NP-complete, heuristics have been proposed to accomplish this task at least approximately. A *heuristic* is a technique designed for solving a problem more quickly when classic methods are too slow, or for finding an approximate solution when classic methods fail to find an exact solution. This is achieved by trading optimality, completeness, accuracy, or precision for speed. Where finding an optimal solution is impossible or impractical, heuristic methods can be used to speed up the process of finding a satisfactory solution. In the context of decision trees, top-down and bottom-up heuristics are employed, see e.g. Rokach and Maimon (2015). *Top-down heuristics* help to successively split the data set at each node by choosing the best available test. By doing so in recursive manner, a decision tree is growing. Subsequently, *bottom-up heuristics* prune the decision tree by removing those dispensable subtrees which do not contribute to optimality as much.

9.2.2.1 Binary Classification

In order to be specific, we consider decision trees for *binary classification*. As above, the classified objects are stored within the set

$$\mathcal{X} = \{x_1, \ldots, x_n\} \subset \mathcal{O},$$

and there are finitely many tests:

$$\mathcal{T} = \{T_1, \ldots, T_m\}.$$

Let any object $x_i \in \mathcal{X}$, $i = 1, \ldots, n$, be assigned to one of the two classes C_{yes} or C_{no}. The misclassification rate of the decision tree D then becomes:

$$\mu(D) = \frac{\#\{i \mid x_i \in C_{yes}, D(x_i) = C_{no}\} + \#\{i \mid x_i \in C_{no}, D(x_i) = C_{yes}\}}{n}.$$

Here, we count those objects from C_{yes} or C_{no}, which are wrongly classified by the decision tree D as from C_{no} or C_{yes}, respectively. Instead of minimizing the misclassification rate directly:

$$\min_D \mu(D),$$

the top-down heuristics rather try to recursively construct a decision tree by an optimal splitting of the data set. The optimality of such a split is understood with respect to a suitably chosen generalization error.

9.2.2.2 Generalization Error

Let us study what happens if we voluntary classify a data subset $S \subset \mathcal{X}$ by putting all its samples either to C_{yes} or C_{no}. This would cause a *generalization error*, which can be expressed in terms of the probability distribution:

$$p_S = \frac{|S \cap C_{yes}|}{|S|}, \quad 1 - p_S = \frac{|S \cap C_{no}|}{|S|},$$

where p_S denotes the share of the data samples in S from the class C_{yes} and $1 - p_S$ denotes the share of the data samples in S from the class C_{no}. Let us present some particular choices for the generalization error $\varepsilon(S)$:

- *Train error* corresponds to the assumption that the classification follows the majority vote. According to the latter, all samples in S are assigned to C_{yes} or C_{no}, depending on where most of them actually belong to. By doing so, we misclassify $\min\{|S \cap C_{yes}|, |S \cap C_{no}|\}$ objects. This motivates to define the train error on the subset S:

$$\varepsilon_1(S) = \min\{p_S, 1 - p_S\}.$$

- *Entropy* corresponds to the assumption that the classification is a random process. According to the latter, an object from S is assigned to C_{yes} or C_{no} with the probability p_S or $1 - p_S$, respectively. Entropy quantifies how surprising the random classification is, averaged on all its possible outcomes:

$$\varepsilon_2(S) = -p_S \cdot \log_2 p_S - (1 - p_S) \cdot \log_2 (1 - p_S).$$

Fig. 9.5 Generalization errors

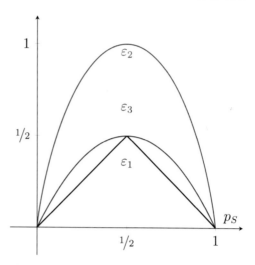

- *Gini index* also corresponds to the assumption that the classification is a random process. According to the latter, an object from S is again assigned to C_{yes} or C_{no} with the probability p_S or $1 - p_S$, respectively. Gini index measures the probability that a data sample is wrongly classified, i.e. it belongs to C_{yes}, but is assigned to C_{no}, or vice versa:

$$\varepsilon_3(S) = 2 \cdot p_S \cdot (1 - p_S).$$

Both the entropy and the Gini index are smooth upper bounds of the train error, see Fig. 9.5.

9.2.2.3 Splitting
Let us consider how a data subset $S \subset \mathcal{X}$ is split by means of a test $T_j \in \mathcal{T}$, $j = 1, \ldots, m$. Such a splitting leads to the subsets of S, whose data samples fail the j-th test or pass through it:

$$L_j = \left\{ x_i \in S \mid T_j(x_i) = 0 \right\}, \quad R_j = \left\{ x_i \in S \mid T_j(x_i) = 1 \right\}.$$

The data samples from L_j will be subsequently classified within the left subtree, and the data samples from R_j within the right subtree. Let us associate with the splitting a gain. Given a generalization error, we define the corresponding *gain* from the splitting by the test T_j as the difference between the generalization error before and after the split:

$$G_j(S) = \varepsilon(S) - \underbrace{\left(\frac{|L_j|}{|S|} \cdot \varepsilon(L_j) + \frac{|R_j|}{|S|} \cdot \varepsilon(R_j) \right)}_{\substack{\text{expected} \\ \text{generalization error}}}.$$

By means of the gain, we are able to compare the tests between each other. As we shall immediately see, this is at the core of top-down heuristics.

9.2.2.4 Iterative Dichotomizer

Let us present a top-down heuristic aiming to construct a decision tree based on a chosen generalization error. Its main idea is to split the data set in such a way that the corresponding gain is maximal. Afterwards, the algorithm proceeds recursively on the left and right subtrees. The *iterative dichotomizer* mainly suggested by Quinlan (1986) has as input a data subset $S \subset \mathcal{X}$ and an index subset of tests $J \subset \{1, \dots, m\}$. As output it returns a decision tree $D = \mathrm{ID}(S, J)$:

Stopping: If all data samples in S are assigned to C_{yes} or C_{no}, then return a leaf with C_{yes} or C_{no}, respectively. If J is empty, then return a leaf according to the majority vote.

Splitting: Else split the data subset $S = L_j \cup R_j$ by the test T_j, so that

$$j \in \arg\max_{j \in J} \ G_j(S).$$

Recursion: Call $D_0 = \mathrm{ID}\left(L_j, J \backslash \{j\}\right)$ for the left and $D_1 = \mathrm{ID}\left(R_j, J \backslash \{j\}\right)$ for the right subtree, and return:

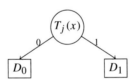

We point out that the iterative dichotomizer falls into the framework of *greedy algorithms* that make the locally optimal choice at each stage, but not necessarily produce an optimal solution. The choice made by a greedy algorithm may depend on choices made so far, but not on future choices. It iteratively performs one greedy step after another, reducing the given problem into a smaller one. In other words, a greedy algorithm never reconsiders its choices. This is why we cannot in general expect that the iterative dichotomizer will arrive at a decision tree D with the minimal misclassification rate $\mu(D)$. The iterative dichotomizer may rather get stuck in a local minimum.

9.2.2.5 Pruning

The iterative dichotomizer usually returns decision trees D of large size, e.g. measured in terms of the average external path length $\rho(D)$, see Exercise 9.4. Such trees tend to have low misclassification rate $\mu(D)$ on the given data set $\mathcal{X}$, but their performance on the real data sets is often poor. This is due to the possible *overfitting* of the training data.

One solution to overcome this drawback is to limit the number of iterations, leading to a decision tree with a bounded number of tests. Another common solution is to prune the already built decision tree. The goal of *pruning* is to reduce it to a much smaller tree, without to considerably worsen the misclassification rate. Let us describe a variant of the pruning procedure from Shalev-Shwartz and Ben-David (2014) as applied for a given decision tree D. The pruning is performed by a bottom-up walk on the decision tree beginning from the leaves of D to its root. Each test node might be replaced by a leaf C_{yes} and C_{no}, or by the left and right subtree of D. By doing so, we obtain the decision trees $D_{yes}, D_{no}, D_{left}, D_{right}$. Their misclassification rates are compared among themselves, as well as with that of D:

$$\min \left\{ \mu\left(D_{yes}\right), \mu\left(D_{no}\right), \mu\left(D_{left}\right), \mu\left(D_{right}\right), \mu(D)\right\}.$$

As a result, we determine the decision tree with the minimal misclassification rate:

$$D' \in \left\{D_{yes}, D_{no}, D_{left}, D_{right}, D\right\},$$

and set $D = D'$. Then, a new test node is considered, and the pruning goes on. The pruning methodology has been developed by Breiman et al. (1984), and since then it is widely used to improve the misclassification rate of decision trees, especially in noisy data domains.

9.3 Case Study: Chess Engine

Chess is one of the most complex board games. The number of possible positions is estimated to be over 10^{43}. Already for the first 40 moves, the estimates are as high as 10^{115} to 10^{120} different courses of play. In game theory, chess falls under the finite zero-sum games with perfect information. Theoretically, it could be determined whether white or black wins if both choose optimal strategies, or whether the games ends up in a draw. However, given the current state of knowledge, it is practically impossible to clarify this question by completely calculating the search tree, due to the large number of positions to be calculated. That's why best *chess engines*, such as e. g. Stockfish, employ sophisticated techniques in order to cut off most of the irrelevant tree branches. Roughly speaking, computer chess playing consists of the following steps:

(1) Construct a sufficiently large search tree with white and black moves as nodes.
(2) Evaluate every position occurring at the leaves.
(3) Determine the best own move by a bottom-up induction from the leaves to the root.

In what follows, we elaborate in detail the mathematics the most chess engines are based on.

Task 1 An *evaluation function* is used to heuristically determine the relative value of chess positions. It has mainly two components:

(a) material count of pieces measured in pawns:

$$
\begin{aligned}
\text{♛} &= \text{♙♙♙♙♙♙♙♙♙} \\
\text{♜} &= \text{♙♙♙♙♙} \\
\text{♝} &= \text{♙♙♙} \\
\text{♞} &= \text{♙♙♙}
\end{aligned}
$$

(b) positional count, such as mobility bonus, number of squares controlled, castled or exposed king, two bishops advantage, control of semi- or open files, rook behind a pawn, passed pawn, enemy checks etc., say, any of these advantages bringing half a pawn.

Resulting value is typically given as a linear combination of the material and positional counts. Try to evaluate the following position for white:

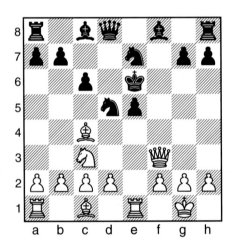

Hint 1 White has given up a knight for a pawn, hence, the material count is -2. However, the white king is castled, whereas black king is exposed; white queen, rook, bishop, and knight are developed, whereas just two black knights are centralized; white rook is on a semi-open file; white has two pawn islands, whereas black has three; white can check, but black cannot. The positional count is thus $0.5 \cdot 5 = 2.5$. Altogether, white is half a pawn ahead:

$$-2 + 2.5 = 0.5.$$

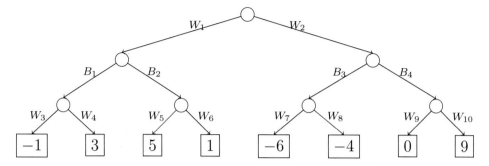

Fig. 9.6 Chess search tree

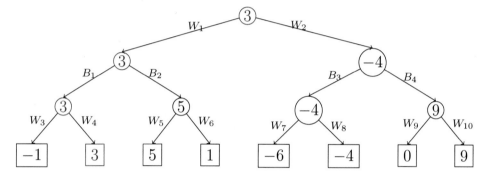

Fig. 9.7 Minimax algorithm

Task 2 Let a search tree consisting of white and black moves be given. The premise of the *minimax algorithm* is that the chess engine will calculate its next best move by evaluating the positions several turns down the road. In doing so, the chess engine chooses the best possible move in order to maximize the evaluation. However, it assumes that the opponent always selects the worst move, i. e. minimizing the evaluation for the computer. Apply the minimax algorithm to the chess search tree in Fig. 9.6. What move will the chess engine play first?

Hint 2 The minimax algorithm applied in a bottom-up manner provides that white would play the move W_1, see Fig. 9.7.

Task 3 How does a chess player decide that a move is advantageous? If during the analysis, one sees that an opponent's reply is unfavourable, then this move will be regarded as refuted. This is the idea behind the so-called alpha-beta pruning. The *alpha-beta pruning* is applied in the situation where we can stop evaluating a part of the search tree if we find a move that leads to a worse situation than a previously discovered move. The alpha-beta pruning does not influence the outcome of the minimax algorithm—it only

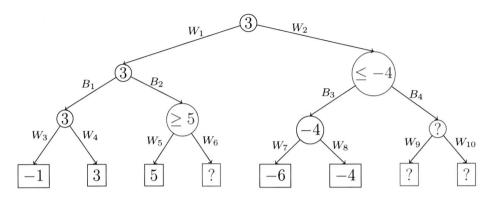

Fig. 9.8 Alpha-beta pruning

makes it faster. Apply the alpha-beta pruning to the chess search tree in Fig. 9.6. How many position evaluations can be saved up?

Hint 3 By applying the alpha-beta pruning, we saved up three out of eight position evaluations, see Fig. 9.8.

Task 4 Try to compare the analysis of the following position by the conventional chess engine Stockfish and the neural network Leela Chess Zero for white:

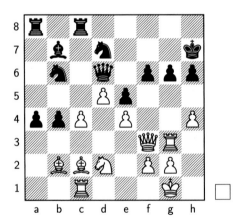

Stockfish	Leela Chess Zero
27 ♖d1 h5 28 ♘f1 ♕e7 29 ♘e3	27 h5 g5 28 ♕f5+ ♔h8 29 f4 a3
♕e8 30 ♘f5 a3 31 ♗c3 bxc3 32	30 ♗a1 ♕e7 31 ♖d1 exf4 32 ♖b3
♘d6 evaluation = 0.19	evaluation = 1.41

Hint 4 Leela Chess Zero achieves a considerable advantage of 1.41 in the final position, whereas Stockfish cannot progress by getting into a nearly equal position with just 0.19. This is due to the first decisive move **27 h5** of Leela Chess Zero. Note that by making the first move **27 ♖d1**, Stockfish just looses time. In fact, black freezes the pawn structure afterwards, and the game remains quiet. Stockfish does not see here the possibility to open up the position by attacking the black king. Note that the competition between Stockfisch and Leela Chess Zero attracts a lot of attention among chess players. Not only that each of them outperforms best human grandmaster, including the current world champion Magnus Carlsen, these and other chess engines also push forward the theory of openings and endings.

9.4   Exercises

Exercise 9.1 (Binary Classification) Let the training set be $\{0, 1\}^3$. It is possible to test any entry of $x \in \{0, 1\}^3$ by the tests

$$
T_j(x) = \begin{cases} 0, \text{ if } x_j = 0, \\ 1, \text{ if } x_j = 1, \end{cases}
$$

where $j = 1, 2, 3$. Find a worst-case binary class assignment, so that the minimal average external path length of the decision trees with zero misclassification error is maximal.

Exercise 9.2 (Minimal External Path Length) Let $f(n)$ denote the minimal external path length of an identification decision tree on an n-element set, where all 1- and 3-element subsets are available as tests. Compute the values of $f(n)$ for $n = 1, \ldots, 8$.

Exercise 9.3 (Matching Problem) Let $\mathcal{A}$ be a finite set and $\mathcal{M} \subset \mathcal{A} \times \mathcal{A} \times \mathcal{A}$. A susbset $\mathcal{M}^* \subset \mathcal{M}$ is called a *matching* if no two elements of $\mathcal{M}^*$ agree in any coordinate, i.e. for all $(a_1, a_2, a_3), (b_1, b_2, b_3) \in \mathcal{M}^*$ it holds:

$$
a_1 \neq b_1, \quad a_2 \neq b_2, \quad a_3 \neq b_3.
$$

The *matching problem* 3DM$(\mathcal{A}, \mathcal{M})$ is to decide whether there exists a matching $\mathcal{M}^* \subset \mathcal{M}$ with $|\mathcal{M}^*| = |\mathcal{A}|$. The matching problem is one of the 21 NP-complete problems provided by Karp (1972). Show that the matching problem 3DM$(\mathcal{A}, \mathcal{M})$ is reducible to the exact cover problem EC3$(\mathcal{Y}, \mathcal{E})$ in polynomial time. Use this result to deduce that the exact cover problem is NP-complete.

Exercise 9.4 (Suboptimality of Iterative Dichotomizer) Let the following classified data set be given:

	Feature 1	Feature 2	Feature 3	Class
Object x_1	1	1	1	C_{yes}
Object x_2	1	1	0	C_{no}
Object x_3	1	0	0	C_{yes}
Object x_4	0	0	1	C_{no}

The features of objects can be independently tested. Construct a decision tree D by applying the iterative dichotomizer with the train error ε_1. Compute the corresponding average external path length $\rho(D)$. Is it minimal among all decision trees with zero misclassification rate?

Exercise 9.5 (Information Gain) Let a data subset $S \subset \mathcal{X}$ be split by means of a test $T \in \mathcal{T}$ into:

$$L = \{x \in S \mid T(x) = 0\}, \quad R = \{x \in S \mid T(x) = 1\}.$$

We denote by X the random variable for the class C_{yes} or C_{no} of an object, i.e. it follows the probability distribution:

$$p_S = \frac{|S \cap C_{yes}|}{|S|}, \quad 1 - p_S = \frac{|S \cap C_{no}|}{|S|}.$$

We denote by Y the random variable for the left subtree L or the right subtree R, where an object belongs to, i.e. it follows the probability distribution:

$$q_S = \frac{|L|}{|S|}, \quad 1 - q_S = \frac{|R|}{|S|}.$$

Show that for the information gain from the splitting by the test T it holds:

$$G(S) = H(X) + H(Y) - H(X, Y),$$

where for a random variable Z with probability distribution $p \in [0, 1]^k$ the entropy is defined as

$$H(Z) = -\sum_{\ell=1}^{k} p_\ell \cdot \log_2 p_\ell.$$

Exercise 9.6 (ID Decision Tree) We consider an infection with fever, cough, and dyspnea as symptoms. Let the following data set of persons, who are known to have been infected or not, be given:

	Fever	Cough	Dyspnea	Infection
Person 1	No	No	No	No
Person 2	Yes	Yes	Yes	Yes
Person 3	Yes	Yes	No	No
Person 4	Yes	No	Yes	Yes
Person 5	Yes	Yes	Yes	Yes
Person 6	No	Yes	No	No
Person 7	Yes	No	Yes	Yes
Person 8	Yes	No	Yes	Yes
Person 9	No	Yes	Yes	Yes
Person 10	Yes	Yes	No	Yes
Person 11	No	Yes	No	No
Person 12	No	Yes	Yes	No
Person 13	No	Yes	Yes	No
Person 14	Yes	Yes	No	No

Construct a decision tree by applying the iterative dichotomizer with the entropy ε_2 as the generalization error. Try to subsequently prune this decision tree by using the misclassification rate as the pruning criterion. What would be predicted for a person without fever and cough, but with dyspnea?

Solutions

10.1 Ranking

Exercise 1.1 (Rankings)
The transition matrix of the network N1 is

$$P = \begin{pmatrix} 0 & 0 & 1 & 1/3 & 1/3 \\ 1 & 0 & 0 & 0 & 1/3 \\ 0 & 1/2 & 0 & 1/3 & 1/3 \\ 0 & 1/2 & 0 & 0 & 0 \\ 0 & 0 & 0 & 1/3 & 0 \end{pmatrix}.$$

The Gaussian elimination for solving $(P - I) \cdot x = 0$ provides:

$$
\begin{array}{ccccc}
-1 & 0 & 1 & 1/3 & 1/3 \\
1 & -1 & 0 & 0 & 1/3 \\
0 & 1/2 & -1 & 1/3 & 1/3 \\
0 & 1/2 & 0 & -1 & 0 \\
0 & 0 & 0 & 1/3 & -1
\end{array}
\quad
\begin{array}{ccccc}
-1 & 0 & 1 & 1/3 & 1/3 \\
0 & -1 & 1 & 1/3 & 2/3 \\
0 & 1/2 & -1 & 1/3 & 1/3 \\
0 & 1/2 & 0 & -1 & 0 \\
0 & 0 & 0 & 1/3 & -1
\end{array}
\sim
\begin{array}{ccccc}
-1 & 0 & 1 & 1/3 & 1/3 \\
0 & -1 & 1 & 1/3 & 2/3 \\
0 & 0 & -1/2 & 1/2 & 2/3 \\
0 & 0 & 1/2 & -5/6 & 1/3 \\
0 & 0 & 0 & 1/3 & -1
\end{array}
$$

$$
\sim
\begin{array}{ccccc}
-1 & 0 & 1 & 1/3 & 1/3 \\
0 & -1 & 1 & 1/3 & 2/3 \\
0 & 0 & -1/2 & 1/2 & 2/3 \\
0 & 0 & 0 & -1/3 & 1 \\
0 & 0 & 0 & 1/3 & -1
\end{array}
\quad
\begin{array}{ccccc}
-1 & 0 & 1 & 1/3 & 1/3 \\
0 & -1 & 1 & 1/3 & 2/3 \\
0 & 0 & -1/2 & 1/2 & 2/3 \\
0 & 0 & 0 & -1/3 & 1 \\
0 & 0 & 0 & 0 & 0
\end{array}
\sim
\begin{array}{ccccc}
1 & 0 & -1 & -1/3 & -1/3 \\
0 & 1 & -1 & -1/3 & -2/3 \\
0 & 0 & 1 & -1 & -4/3 \\
0 & 0 & 0 & 1 & -3 \\
0 & 0 & 0 & 0 & 0
\end{array}
$$

© Springer-Verlag GmbH Germany, part of Springer Nature 2021
V. Shikhman, D. Müller, *Mathematical Foundations of Big Data Analytics*,
https://doi.org/10.1007/978-3-662-62521-7_10

$$
\begin{array}{ccccc}
1 & 0 & -1 & 0 & -4/3 \\
0 & 1 & -1 & 0 & -5/3 \\
\sim\; 0 & 0 & 1 & 0 & -13/3 \\
0 & 0 & 0 & 1 & -3 \\
0 & 0 & 0 & 0 & 0
\end{array}
\qquad
\begin{array}{ccccc}
1 & 0 & 0 & 0 & -17/3 \\
0 & 1 & 0 & 0 & -18/3 \\
\sim\; 0 & 0 & 1 & 0 & -13/3 \\
0 & 0 & 0 & 1 & -9/3 \\
0 & 0 & 0 & 0 & 0
\end{array}
$$

The solutions can be written by means of a free parameter $t \in \mathbb{R}$ as

$$
x = t \cdot
\begin{pmatrix}
17/3 \\
18/3 \\
13/3 \\
9/3 \\
1
\end{pmatrix}.
$$

We may choose $t > 0$ to guarantee that the components of x are positive. In order to enforce that they sum up to one, we specify the free parameter:

$$
1 = e^T \cdot x = t \cdot \left(\frac{17}{3} + \frac{18}{3} + \frac{13}{3} + \frac{9}{3} + 1 \right) = t \cdot \frac{60}{3} \quad \Rightarrow \quad t = \frac{3}{60}.
$$

From here we get the ranking

$$
x = \frac{3}{60} \cdot
\begin{pmatrix}
17/3 \\
18/3 \\
13/3 \\
9/3 \\
1
\end{pmatrix}
=
\begin{pmatrix}
17/60 \\
18/60 \\
13/60 \\
9/60 \\
3/60
\end{pmatrix}.
$$

The Google ranking for the network N1 corresponds to the eigenvector of the regularized matrix:

$$
P_\alpha = (1 - \alpha) \cdot P + \alpha \cdot E,
$$

where $\alpha = 0.15$ and E is the stochastic (5×5)-matrix with the entries equal to $1/5$. For computation of the Google ranking for the network N1 we refer to the Python code.

Exercise 1.2 (Cesàro Mean)

(i) The Cesàro means have nonnegative components, since

$$\bar{x}(s) = \underbrace{\frac{1}{s}}_{\geq 0} \cdot \sum_{t=1}^{s} \underbrace{x(t)}_{\geq 0} \geq 0.$$

Besides, their components sum up to one:

$$e^T \cdot \bar{x}(s) = e^T \cdot \frac{1}{s} \cdot \sum_{t=1}^{s} x(t) = \frac{1}{s} \cdot \sum_{t=1}^{s} \underbrace{e^T \cdot x(t)}_{=1} = \frac{1}{s} \cdot \sum_{t=1}^{s} 1 = 1.$$

Both assertions show that $\bar{x}(s)$ is a distribution for $s = 1, 2, \ldots$

(ii) Due to the formula

$$x(t+1) = P \cdot x(t) = P \cdot P \cdot x(t-1) = P^2 \cdot x(t-1) = \ldots = P^t \cdot x(1),$$

we have the following representation of the Cesàro means:

$$\bar{x}(s) = \frac{1}{s} \cdot \sum_{t=1}^{s} x(t) = \frac{1}{s} \cdot \sum_{t=1}^{s} P^{t-1} \cdot x(1).$$

Using the triangle inequality, we estimate:

$$\|\bar{x}(s) - P \cdot \bar{x}(s)\|_1 = \left\| \frac{1}{s} \cdot \sum_{t=1}^{s} P^{t-1} \cdot x(1) - P \cdot \frac{1}{s} \cdot \sum_{t=1}^{s} P^{t-1} \cdot x(1) \right\|_1$$

$$= \frac{1}{s} \cdot \left\| \sum_{t=1}^{s} P^{t-1} \cdot x(1) - \sum_{t=1}^{s} P^{t} \cdot x(1) \right\|_1 = \frac{1}{s} \cdot \left\| x(1) - P^s \cdot x(1) \right\|_1$$

$$\leq \frac{1}{s} \cdot \left(\|x(1)\|_1 + \left\| P^s \cdot x(1) \right\|_1 \right) = \frac{1}{s} \cdot \left(\|x(1)\|_1 + \|x(s+1)\|_1 \right)$$

$$= \frac{1}{s} \cdot \left(e^T \cdot \underbrace{|x(1)|}_{\geq 0} + e^T \cdot \underbrace{|x(s+1)|}_{\geq 0} \right)$$

$$= \frac{1}{s} \cdot \left(\underbrace{e^T \cdot x(1)}_{=1} + \underbrace{e^T \cdot x(s+1)}_{=1} \right) = \frac{2}{s}.$$

The sequence of Cesàro means approaches the set of rankings due to

$$\|\bar{x}(s) - P \cdot \bar{x}(s)\|_1 \leq \frac{2}{s} \to 0 \quad \text{for } s \to \infty.$$

Exercise 1.3 (Permutation Matrices)

(i) The transition matrix is

$$P = \begin{pmatrix}
 & \boxed{1} & \boxed{2} & \cdots & \boxed{n-1} & \boxed{n} \\
\boxed{1} & 0 & 0 & \cdots & 0 & 1 \\
\boxed{2} & 1 & 0 & \cdots & 0 & 0 \\
\boxed{3} & 0 & 1 & \cdots & 0 & 0 \\
 & \vdots & \vdots & \vdots & \cdots & \vdots & \vdots \\
\boxed{n} & 0 & 0 & \cdots & 1 & 0
\end{pmatrix}.$$

(ii) We solve the system $(P - I) \cdot x = 0$ of linear equations with

$$P - I = \begin{pmatrix}
-1 & 0 & \cdots & 0 & 1 \\
1 & -1 & \cdots & 0 & 0 \\
0 & 1 & \cdots & 0 & 0 \\
\vdots & \vdots & \cdots & \vdots & \vdots \\
0 & 0 & \cdots & 1 & -1
\end{pmatrix}.$$

For the components of x it follows:

$$x_n = x_1, \quad x_1 = x_2, \quad \ldots, \quad x_{n-1} = x_n.$$

From here we see that they are equal to each other. Hence, the unique ranking x is given by the uniform distribution

$$x_1 = x_2 = \ldots = x_n = \frac{1}{n}.$$

(iii) Let the iteration scheme ($\mathcal{I}$) start from an arbitrary distribution

$$x(1) = (a_1, a_2, a_3, \ldots, a_n)^T.$$

From $x(t + 1) = P \cdot x(t)$ we obtain:

$$x(2) = \begin{pmatrix} a_n \\ a_1 \\ \vdots \\ a_{n-2} \\ a_{n-1} \end{pmatrix}, \quad x(3) = \begin{pmatrix} a_{n-1} \\ a_n \\ \vdots \\ a_{n-3} \\ a_{n-2} \end{pmatrix}, \quad \ldots, \quad x(n + 1) = \begin{pmatrix} a_1 \\ a_2 \\ \vdots \\ a_{n-1} \\ a_n \end{pmatrix}.$$

In each iteration the components are shifted at exactly one index further. Therefore, after n iterations the process is arriving at the starting distribution, i.e. $x(n + 1) = x(1)$, which means that it oscillates.

(iv) From Exercise 1.2 (ii) we know that the sequence of Cesàro means approaches the set of rankings. Since the set of rankings is a singleton due to (ii), the sequence of Cesàro means has the ranking x as its limit.

Exercise 1.4 (Positive Matrices)

(i) Assume that for the smallest entry of P holds:

$$\min_{1 \le i, j \le n} p_{ij} > \frac{1}{n}.$$

Then, at least one component of $e^T \cdot P$ strictly exceeds 1. This contradicts to the fact that P is a stochastic matrix.

(ii) If $\bar{\alpha} = 1$, then $P = E$ provides the desired representation. For $\bar{\alpha} < 1$ we consider the matrix

$$\bar{P} = \frac{1}{1 - \bar{\alpha}} \cdot (P - \bar{\alpha} \cdot E).$$

It is straightforward to see that $\bar{P}$ is stochastic:

$$\bar{p}_{ij} = \frac{1}{1 - \bar{\alpha}} \cdot \left(p_{ij} - \frac{\bar{\alpha}}{n} \right) = \frac{1}{1 - \bar{\alpha}} \cdot \left(p_{ij} - \min_{1 \le i, j \le n} p_{ij} \right) \ge 0,$$

$$e^T \cdot \bar{P} = \frac{1}{1 - \bar{\alpha}} \cdot \left(e^T \cdot P - \bar{\alpha} \cdot e^T \cdot E \right) = \frac{1}{1 - \bar{\alpha}} \cdot \left(e^T - \bar{\alpha} \cdot e^T \right) = e^T.$$

(iii) We estimate for any ranking x:

$$\|x(t+1) - x\|_1 = \|P \cdot (x(t) - x)\|_1 = \left\|\left((1-\bar{\alpha}) \cdot \bar{P} + \bar{\alpha} \cdot E\right) \cdot (x(t) - x)\right\|_1$$

$$= \left\|(1-\bar{\alpha}) \cdot \bar{P} \cdot (x(t) - x) + \bar{\alpha} \cdot \underbrace{(E \cdot x(t) - E \cdot x)}_{=e-e=0}\right\|_1$$

$$= (1-\bar{\alpha}) \cdot \left\|\bar{P} \cdot (x(t) - x)\right\|_1 = (1-\bar{\alpha}) \cdot e^T \cdot \underbrace{\left|\bar{P} \cdot (x(t) - x)\right|}_{\leq \bar{P} \cdot |x(t) - x|}$$

$$\leq (1-\bar{\alpha}) \cdot \underbrace{e^T \cdot \bar{P}}_{=e^T} \cdot |x(t) - x| = (1-\bar{\alpha}) \cdot \|x(t) - x\|_1 .$$

(iv) Due to $0 < \bar{\alpha} \leq 1$, we recursively have:

$$\|x(t+1) - x\|_1 \leq \leq (1-\bar{\alpha})^t \cdot \|x(1) - x\|_1 \to 0 \quad \text{for } t \to \infty.$$

Hence, the convergence rate of $\mathcal{I}$ is $1 - \bar{\alpha}$, and x is the unique ranking.

Exercise 1.5 (Social Status)
In the Facebook network, the columns of the transition matrix are given by the shares of granted likes:

$$P = \begin{pmatrix} & \boxed{1} & \boxed{2} & \boxed{3} & \boxed{4} \\ \boxed{1} & 0 & 0 & 2/8 & 1/3 \\ \boxed{2} & 5/9 & 0 & 2/8 & 1/3 \\ \boxed{3} & 0 & 7/9 & 0 & 1/3 \\ \boxed{4} & 4/9 & 2/9 & 4/8 & 0 \end{pmatrix} .$$

This corresponds to the attention paid by friends to each other. The ranking corresponding to the transition matrix P can be easily computed:

$$x = \begin{pmatrix} 405/2429 \\ 630/2429 \\ 716/2429 \\ 678/2429 \end{pmatrix} .$$

Friend $\boxed{3}$ has the highest social status, although receiving in total just eight likes. Whereas friend $\boxed{4}$ receives ten likes, but has the lower social status. For the latter it is crucial to get likes from influential friends rather than to maximize their number.

Exercise 1.6 (Exchange Economy, See Gale (1960))
We start with defining the exchange matrix $A = (a_{ij})$, where a_{ij} is the amount of the good G_j employed by the producer P_i while manufacturing:

$$
\begin{pmatrix}
 & \boxed{G_1} \cdots \boxed{G_j} \cdots \boxed{G_n} \\
\boxed{P_1} & a_{11} \cdots a_{1j} \cdots a_{1n} \\
 & \vdots \quad \vdots \quad \vdots \quad \vdots \\
\boxed{P_i} & a_{i1} \cdots a_{ij} \cdots a_{in} \\
 & \vdots \quad \vdots \quad \vdots \quad \vdots \\
\boxed{P_n} & a_{n1} \cdots a_{nj} \cdots a_{nn}
\end{pmatrix}.
$$

The i-th row of A contains the amounts of goods $G_1, \ldots, G_n$ employed by the producer P_i in order to manufacture one entity of the good G_i. The j-th column of A contains the amounts of the good G_j employed by producers $P_1, \ldots, P_n$. As just an entity of the good G_j is manufactured and can be thus distributed among the producers $P_1, \ldots, P_n$, it holds:

$$
\sum_{i=1}^{n} a_{ij} = 1 \quad \text{for } j = 1, \ldots, n.
$$

Equivalently, the exchange matrix is stochastic, i.e.

$$
e^T = e^T \cdot A.
$$

The equilibrium prices $p = (p_1, \ldots, p_n)^T \geq 0$ are characterized by the following inequalities:

$$
p_i \geq \sum_{j=1}^{n} a_{ij} \cdot p_j \quad \text{for } i = 1, \ldots, n.
$$

The i-th inequality says that the revenue of the producer P_i from selling one entity of the good G_i has to be greater or equal than the cost from its manufacturing. In matrix form we have for that:

$$
p \geq A \cdot p.
$$

Here, actually the equality holds. To prove this, we notice first that

$$e^T \cdot (p - A \cdot p) = e^T \cdot p - \underbrace{e^T \cdot A}_{=e^T} \cdot p = e^T \cdot p - e^T \cdot p = 0.$$

It follows:

$$\underbrace{e^T}_{>0} \cdot \underbrace{(p - A \cdot p)}_{\geq 0} = 0 \quad \Rightarrow \quad p = A \cdot p.$$

The vector of equilibrium prices is thus a ranking and we can suggest to use the iteration scheme $(\mathcal{I})$ for price adjustment. The prices in the following period are based on the manufacturing costs of the recent period, i.e.

$$p(t+1) = A \cdot p(t) \quad \text{for } t = 1, 2, \ldots,$$

where $p(1)$ is a normalized vector of starting prices.

10.2   Online Learning

Exercise 2.1 (Dual Norm)
We show that the dual norm satisfies the properties of a norm, i.e. positive definiteness, absolute homogeneity, and triangle inequality.

(i) Clearly, it holds:

$$\|0\|_* = \max_{\|x\| \leq 1} 0^T \cdot x = 0.$$

Next, take an arbitrary $g \neq 0$, which implies $\|g\| \neq 0$. Hence, we can define a feasible vector by choosing $x = \frac{g}{\|g\|}$ and calculate:

$$\|g\|_* = \max_{\|x\| \leq 1} g^T \cdot x \geq g^T \cdot \frac{g}{\|g\|} = \frac{\|g\|_2^2}{\|g\|} > 0.$$

Thus, we conclude that $\|g\|_* = 0$ if and only if $g = 0$, and positive definiteness holds.

(ii) Absolute homogeneity is satisfied, since by symmetry of the feasible set we have:

$$\max_{\|x\| \leq 1} g^T \cdot x = \max_{\|x\| \leq 1} |g^T \cdot x|.$$

Therefore, for any $\alpha \in \mathbb{R}$ and $g \in \mathbb{R}^n$ it holds:

$$\|\alpha \cdot g\|_* = \max_{\|x\| \leq 1} |\alpha \cdot g^T \cdot x| = |\alpha| \cdot \max_{\|x\| \leq 1} g^T \cdot x = |\alpha| \cdot \|g\|_*.$$

(iii) Take arbitrary $g, h \in \mathbb{R}^n$ and consider the dual norm of their sum:

$$\|g + h\|_* = \max_{\|x\| \leq 1} \left(g^T + h^T\right) \cdot x \leq \max_{\|x\| \leq 1} g^T \cdot x + \max_{\|x\| \leq 1} h^T \cdot x = \|g\|_* + \|h\|_*.$$

Hence, the triangle inequality is fulfilled.

Exercise 2.2 (Cauchy-Schwarz Inequality)
Let us show that the Euclidean norm is self-dual:

$$\|g\|_* = \max_{\|x\|_2 \leq 1} g^T \cdot x = \|g\|_2.$$

The assertion trivial holds if $g = 0$. It remains to analyze the case $g \neq 0$. Note that the problem consists of maximizing a linear function on a convex set, which means that the maximum is attained at the boundary of the feasible set, i.e. we solve

$$\max_{\|x\|_2 = 1} g^T \cdot x.$$

By reformulating the equality constraint in the equivalent form $\frac{1}{2}\|x\|_2^2 = \frac{1}{2}$ and introducing the Lagrange multiplier $\mu \in \mathbb{R}^n$ for the equality constraint, we apply the Lagrange multiplier rule, see e.g. Jongen et al. (2004):

$$g - \mu \cdot x = 0.$$

Multiplying by x^T from the left and using $x^T \cdot x = 1$, we see that

$$g^T \cdot x = \mu,$$

i.e. μ is the optimal value of the objective function. Hence, for the maximizer we get $x = \frac{1}{\mu} \cdot g$. Plugging this into the equality constraint leads to

$$\left(\frac{1}{\mu}\right)^2 \cdot g^T \cdot g = 1 \quad \text{or} \quad \|g\|_2^2 = \mu^2.$$

Taking the square root proves the assertion:

$$\|g\|_* = \mu = \|g\|_2.$$

The Cauchy-Schwarz inequality follows directly from the Hölder inequality.

Exercise 2.3 (Three-Points Identity)
Plugging the definition of the Bregman divergence into the left-hand side provides:

$$B(x, y) - B(x, z) - B(z, y) = d(x) - d(y) - \nabla^T d(y) \cdot (x - y)$$

$$- \left(d(x) - d(z) - \nabla^T d(z) \cdot (x - z) \right)$$

$$- \left(d(z) - d(y) - \nabla^T d(y) \cdot (z - y) \right)$$

$$= -\nabla^T d(y) \cdot (x - y) + \nabla^T d(z) \cdot (x - z) + \nabla^T d(y) \cdot (z - y)$$

$$= \nabla^T d(y) \cdot (-x + y + z - y) + \nabla^T d(z) \cdot (x - z)$$

$$= \left(\nabla^T d(y) - \nabla^T d(z) \right) \cdot (z - x).$$

Exercise 2.4 (Negative Entropy)
Let us first calculate the prox-center $x(1)$ of the negative entropy on the simplex. For that, we solve the optimization problem

$$\min_{x \geq 0} d(x) = \sum_{i=1}^{n} x_i \cdot \ln x_i \quad \text{s.t.} \quad e^T \cdot x - 1 = 0.$$

Introducing the Lagrange multiplier μ for the equality constraint, we get the optimality condition, see e.g. Jongen et al. (2004):

$$\nabla d(x) = \mu \cdot \nabla \left(e^T \cdot x - 1 \right).$$

Written componentwise, the latter provides:

$$x_i = e^{\mu - 1}, \quad i = 1, \ldots, n.$$

Hence, all components of x are equal, from where we conclude that the prox-center is given by

$$x(1) = \left(\frac{1}{n}, \ldots, \frac{1}{n} \right)^T.$$

The negative entropy is, thus, lower bounded for all $y \in \Delta$:

$$d(y) \geq d(x(1)) = \sum_{i=1}^{n} \frac{1}{n} \cdot \ln \frac{1}{n} = \ln \frac{1}{n} = -\ln n.$$

Since $x_i \in [0, 1]$ for all $i = 1, \ldots, n$, the negative entropy is upper bounded by zero:

$$d(x) = \sum_{i=1}^{n} \underbrace{x_i}_{\geq 0} \cdot \underbrace{\ln x_i}_{\leq 0} \leq 0.$$

For the diameter of the simplex Δ we finally get:

$$D^2 = \max_{x, y \in \Delta} d(x) - d(y) \leq 0 + \ln n = \ln n.$$

Exercise 2.5 (Euclidean Setup)

(i) The function $d(x) = \frac{1}{2} \cdot \|x\|_2^2$ is twice continuously differentiable with gradient $\nabla d(x) = x$. Thus, its Hesse matrix is the identity, i.e. $\nabla^2 d(x) = I$. The second order strong-convexity criterion $(\mathcal{SC}^2)$ yields for all $\xi \in \mathbb{R}^n$:

$$\xi^T \cdot I \cdot \xi = \xi^T \cdot \xi = 1 \cdot \|\xi\|_2^2,$$

which shows that $\frac{1}{2} \cdot \|x\|_2^2$ is indeed a prox-function with respect to the Euclidean norm $\|\cdot\|_2$ with the convexity parameter $\beta = 1$.

(ii) We plug $d(x) = \frac{1}{2} \|x\|_2^2$ into the definition of Bregman divergence:

$$B(x, y) = d(x) - d(y) - \nabla^T d(y) \cdot (x - y) = \frac{1}{2} \cdot \|x\|_2^2 - \frac{1}{2} \cdot \|y\|_2^2 - y^T \cdot (x - y)$$

$$= \frac{1}{2} \cdot \|x\|_2^2 + \frac{1}{2} \cdot \|y\|_2^2 - y^T \cdot x = \frac{1}{2} \cdot \left(\|x\|_2^2 + \|y\|_2^2 - 2 \cdot y^T \cdot x \right) = \frac{1}{2} \cdot \|x - y\|_2^2.$$

Exercise 2.6 (Projection)

Note that both optimization problems are strongly convex. Equivalence can therefore be shown by coincidence of their unique minimizers. We start with the auxiliary optimization problem $(\mathcal{A})$:

$$\min_{x \in X} c^T \cdot x + \frac{1}{2} \cdot \|x - y\|_2^2.$$

Adding terms which are independent of the decision variable does not change the solution, so we add $\frac{1}{2} \cdot \|c\|_2^2 - c^T \cdot y$. This proves the equivalence:

$$\underset{x \in X}{\arg\min} \; c^T \cdot x + \frac{1}{2} \cdot \|x - y\|_2^2 = \underset{x \in X}{\arg\min} \; c^T \cdot x + \frac{1}{2} \cdot \|x - y\|_2^2 + \frac{1}{2} \cdot \|c\|_2^2 - c^T \cdot y$$

$$= \underset{x \in X}{\arg\min} \; \frac{1}{2} \cdot \|x - y\|_2^2 + \frac{1}{2} \cdot \|c\|_2^2 + c^T \cdot (x - y)$$

$$= \underset{x \in X}{\arg\min} \; \frac{1}{2} \cdot \left(\|x - y\|_2^2 + \|c\|_2^2 + 2 \cdot c^T \cdot (x - y) \right)$$

$$= \underset{x \in X}{\arg\min} \; \frac{1}{2} \cdot \|x - (y - c)\|_2^2 .$$

Recall that taking square root is a monotone transformation, thus, the assertion holds.

Exercise 2.7 (Online Gradient Descent, See Zinkevich (2003))
The online mirror descent boils down to the online gradient descent. In order to prove this claim, plug the Euclidean divergence into the $(\mathcal{OMD})$ update:

$$x(t + 1) = \underset{x \in X}{\arg\min} \; f_t(x(t)) + \nabla^T f_t(x(t)) \cdot x + \frac{1}{2\eta} \cdot \|x - x(t)\|_2^2 .$$

The latter can be equivalently written as

$$x(t + 1) = \underset{x \in X}{\arg\min} \; \eta \cdot \nabla^T f_t(x(t)) \cdot x + \frac{1}{2} \cdot \|x - x(t)\|_2^2 .$$

But this is exactly in the form of the auxiliary problem $(\mathcal{A})$ with $c = \eta \cdot \nabla f_t(x(t))$. According to Exercise 2.6, this is equivalent to

$$x(t + 1) = \underset{x \in X}{\arg\min} \; \frac{1}{2} \cdot \|x - (x(t) - \eta \cdot \nabla f_t(x(t)))\|_2 .$$

Referring to Exercise 2.6 once more, the unique solution of the latter is

$$x(t + 1) = \text{proj}_X(x(t) - \eta \cdot \nabla f_t(x(t))) .$$

Furthermore, $x(1)$ is the prox-center of the Euclidean prox-function, i.e.

$$x(1) = \underset{x \in X}{\arg\min} \; \frac{1}{2} \cdot \|x\|_2^2 = \underset{x \in X}{\arg\min} \; \frac{1}{2} \cdot \|x - 0\|_2^2 = \text{proj}_X(0) .$$

Overall, the following variant of $(\mathcal{OGD})$ has been derived:

$$x(t+1) = \arg\min_{x \in X} \ \text{proj}_X \left(x(t) - \eta \cdot \nabla f_t\left(x(t)\right)\right), \quad x(1) = \text{proj}_X(0).$$

We turn our attention to the step size parameter η. We recall from Step 2 in the convergence analysis that for any $T > 0$ the regret $\mathcal{R}(T)$ of $(\mathcal{OMD})$ can be bounded above:

$$\mathcal{R}(T) \leq \frac{1}{\eta} \cdot \frac{D^2}{T} + \eta \cdot \frac{G^2}{2\beta}.$$

In order to derive the optimal step size, we minimize the upper bound w.r.t. η:

$$\min_{\eta > 0} \ \frac{1}{\eta} \cdot \frac{D^2}{T} + \eta \cdot \frac{G^2}{2\beta}.$$

The first order condition provides:

$$\frac{1}{\eta^2} \cdot \frac{D^2}{T} = \frac{G^2}{2\beta}.$$

Thus, the optimal step size is given by

$$\eta = \frac{D}{G} \cdot \sqrt{\frac{2\beta}{T}}.$$

With this optimal choice the upper bound for the regret becomes

$$\mathcal{R}(T) \leq D \cdot G \cdot \sqrt{\frac{2}{\beta \cdot T}}.$$

Due to Exercise 2.5, the Euclidean prox-function is 1-strongly convex with respect to the Euclidean norm, and the optimal rate of convergence for the average regret in the Euclidean setup is

$$\mathcal{R}(T) \leq D \cdot G \cdot \sqrt{\frac{2}{T}}.$$

10.3   Recommendation Systems

Exercise 3.1 (User and Movie Similarity)
We want to complete the Netflix rating matrix:

$$R = \begin{pmatrix} & \boxed{M1} & \boxed{M2} & \boxed{M3} & \boxed{M4} \\ \boxed{U1} & 5 & 3 & - & 1 \\ \boxed{U2} & 4 & - & - & 1 \\ \boxed{U3} & 1 & 1 & - & 5 \\ \boxed{U4} & 1 & - & - & 4 \\ \boxed{U5} & - & 1 & 5 & 4 \end{pmatrix}.$$

For that, we first apply the $(k\mathcal{N}\mathcal{N})$ algorithm based on the cosine similarity for users. For users U1 and U2 we have $M_1 \cap M_2 = \{1, 4\}$, i.e. they both rated movies M1 and M4. Then, we calculate:

$$\text{Cosine}(1, 2) = \frac{5 \cdot 4 + 1 \cdot 1}{\sqrt{5^2 + 1^2} \cdot \sqrt{4^2 + 1^2}} \approx 0.99.$$

For users U1 and U3 we have $M_1 \cap M_3 = \{1, 2, 4\}$ and, hence, we get:

$$\text{Cosine}(1, 3) = \frac{5 \cdot 1 + 3 \cdot 1 + 1 \cdot 5}{\sqrt{5^2 + 3^2 + 1^2} \cdot \sqrt{1^2 + 1^2 + 5^2}} \approx 0.42.$$

The similarity of users U1 and U4 is based on the set of movies $M_1 \cap M_4 = \{1, 4\}$, which provides:

$$\text{Cosine}(1, 4) = \frac{5 \cdot 1 + 1 \cdot 4}{\sqrt{5^2 + 1^2} \cdot \sqrt{1^2 + 4^2}} \approx 0.43.$$

It remains to analyze the case of users U1 and U5 with $M_1 \cap M_5 = \{2, 4\}$. This leads to

$$\text{Cosine}(1, 5) = \frac{3 \cdot 1 + 1 \cdot 4}{\sqrt{3^2 + 1^2} \cdot \sqrt{1^2 + 4^2}} \approx 0.54.$$

The other similarities can be derived in the same way:

$$\text{Cosine}(2, 3) \approx 0.43, \quad \text{Cosine}(2, 4) \approx 0.47, \quad \text{Cosine}(2, 5) = 1,$$

$$\text{Cosine}(3, 4) \approx 0.99, \quad \text{Cosine}(3, 5) \approx 0.99, \quad \text{Cosine}(4, 5) = 1.$$

Note that the perfect cosine similarity between users U5 and U2, as well as U4 is due to the fact that they mutually rated only one movie. In order to predict the missing entries, we must find the two nearest neighbors of each user. We show this in detail for user U5. Our goal is to make a recommendation for user U5 on movie M1. The set of users who specified ratings of movie M1 is

$$U_1 = \{1, 2, 3, 4\},$$

from where we are able to select the two nearest neighbors of user U5:

$$N_1(5) = \{2, 4\}.$$

As a last step we calculate the weighted average:

$$r_{51} = \frac{1 \cdot 4 + 1 \cdot 1}{|1| + |1|} = 2.50.$$

The recommendation on movie M3 illustrates the drawback of this algorithm. As only user U5 specified the rating of movie M3, the set $U_3 = \{5\}$ is a singleton. Clearly, the cosine similarities in the weighted average cancel out and the rating of user U5 is set for all other users as

$$r_{13} = r_{23} = r_{33} = r_{43} = 5.00.$$

It remains to predict the ratings r_{22} and r_{42}. The corresponding two nearest neighbors are:

$$N_2(2) = \{1, 5\}, \quad N_2(4) = \{3, 5\}.$$

Therefore, the predicted ratings are:

$$r_{22} = \frac{0.99 \cdot 3 + 1 \cdot 1}{|0.99| + |1|} \approx 1.99, \quad r_{42} = \frac{0.99 \cdot 1 + 1 \cdot 1}{|0.99| + |1|} = 1.$$

This gives the user-based completion matrix:

$$R_{neighbor} = \begin{pmatrix} & \boxed{M1} & \boxed{M2} & \boxed{M3} & \boxed{M4} \\ \boxed{U1} & 5 & 3 & 5.00 & 1 \\ \boxed{U2} & 4 & 1.99 & 5.00 & 1 \\ \boxed{U3} & 1 & 1 & 5.00 & 5 \\ \boxed{U4} & 1 & 1 & 5.00 & 4 \\ \boxed{U5} & 2.50 & 1 & 5 & 4 \end{pmatrix}.$$

Another option to complete the matrix R is to apply an item-based (kNN) algorithm. Instead of following the idea of like-minded users, the similarity between items becomes crucial. This reflects the idea that a user will rate similar movies in a similar manner. The cosine similarity between two movies j and k is

$$\text{Cosine}(j, k) = \frac{\displaystyle\sum_{i \in U_j \cap U_k} r_{ij} \cdot r_{ik}}{\sqrt{\displaystyle\sum_{i \in U_j \cap U_k} r_{ij}^2} \cdot \sqrt{\displaystyle\sum_{i \in U_j \cap U_k} r_{ik}^2}}.$$

Compared to the user-based case where the summation is along columns, we sum along the rows. Note that the summation is over all users who specified a rating for the corresponding two movies. Let us calculate the cosine similarities of movies:

$$\text{Cosine}(1, 2) = \frac{5 \cdot 3 + 1 \cdot 1}{\sqrt{5^2 + 1^2} \cdot \sqrt{3^2 + 1^2}} \approx 0.99,$$

$$\text{Cosine}(1, 4) = \frac{5 \cdot 1 + 4 \cdot 1 + 1 \cdot 5 + 1 \cdot 4}{\sqrt{5^2 + 4^2 + 1^2 + 1^2} \cdot \sqrt{1^2 + 1^2 + 5^2 + 4^2}} \approx 0.42,$$

$$\text{Cosine}(2, 4) = \frac{3 \cdot 1 + 1 \cdot 5 + 1 \cdot 4}{\sqrt{3^2 + 1^2 + 1^2} \cdot \sqrt{1^2 + 5^2 + 4^2}} \approx 0.56,$$

$$\text{Cosine}(2, 3) = \frac{1 \cdot 5}{\sqrt{1^2} \cdot \sqrt{5^2}} = 1,$$

$$\text{Cosine}(3, 4) = \frac{5 \cdot 4}{\sqrt{5^2} \cdot \sqrt{4^2}} = 1.$$

Again the high similarity of movie M3 with M2 and M4, respectively, is due to the fact that only one user U5 rated them mutually. Even though user U5 specified different ratings for movies M2 and M4, they share the same similarity with movie M3. Note that due to $U_1 \cap U_3 = \emptyset$, the cosine similarity of movies M1 and M3 can not be calculated. The item-based ($k\mathcal{NN}$) algorithm is similar to its user-based variant. In order to predict rating r_{ij}, we sort the set of movies with ratings specified by user i in decreasing order with respect to their similarities to movie j. Then, two neighbouring movies are identified and their ratings are averaged by the weighted cosine similarity. We state the results of the predictions:

$$r_{51} = \frac{0.99 \cdot 1 + 0.42 \cdot 4}{|0.99| + |0.42|} \approx 1.89, \quad r_{13} = \frac{1 \cdot 3 + 1 \cdot 1}{|1| + |1|} = 2.00,$$

$$r_{22} = \frac{0.99 \cdot 4 + 0.56 \cdot 1}{|0.99| + |0.56|} \approx 2.92, \quad r_{42} = \frac{0.99 \cdot 1 + 0.56 \cdot 4}{|0.99| + |0.56|} \approx 2.08,$$

$$r_{33} = \frac{1 \cdot 1 + 1 \cdot 5}{|1| + |1|} = 3.00, \qquad r_{23} = \frac{1 \cdot 1}{|1|} = 1.00, \quad r_{43} = \frac{1 \cdot 4}{|1|} = 4.00.$$

The movie-based completion of the matrix R looks as follows:

$$R_{neighbor} = \begin{pmatrix} & \boxed{M1} & \boxed{M2} & \boxed{M3} & \boxed{M4} \\ \boxed{U1} & 5 & 3 & 2.00 & 1 \\ \boxed{U2} & 4 & 2.92 & 1.00 & 1 \\ \boxed{U3} & 1 & 1 & 3.00 & 5 \\ \boxed{U4} & 1 & 2.08 & 4.00 & 4 \\ \boxed{U5} & 1.89 & 1 & 5 & 4 \end{pmatrix}.$$

Exercise 3.2 (Eigenvalues and Singular Values)
By using the singular value decomposition of the matrix $R = U \cdot \Sigma \cdot V$, we obtain:

$$R \cdot R^T \cdot U = U \cdot \Sigma \cdot \underbrace{V \cdot V^T}_{=I} \cdot \Sigma^T \cdot \underbrace{U^T \cdot U}_{=I} = U \cdot \Sigma^2.$$

Rewritten in vector form, this means:

$$R \cdot R^T \cdot u_i = \sigma_i^2 \cdot u_i \quad \text{for all } i = 1, \dots, r.$$

Hence, $\lambda_i = \sigma_i^2$ are positive eigenvalues of $R \cdot R^T$ for all $i = 1, \dots, r$. Assume that there is another eigenvalue $\lambda \neq 0$ of $R \cdot R^T$ with a corresponding eigenvector u, i.e.

$$R \cdot R^T \cdot u = \lambda \cdot u.$$

Then, we have:

$$\lambda \cdot u_i^T \cdot u = u_i^T \cdot \underbrace{\lambda \cdot u}_{=R \cdot R^T \cdot u} = u_i^T \cdot R \cdot R^T \cdot u = \underbrace{\left(R \cdot R^T \cdot u_i \right)^T}_{=\lambda_i \cdot u_i^T} \cdot u = \lambda_i \cdot u_i^T \cdot u.$$

Since $\lambda \neq \lambda_i$, we deduce from here that $u_i^T \cdot u = 0$, hence, u is orthogonal to u_i, $i = 1, \dots, r$. Moreover, u lies in the range of R:

$$R \cdot \left(\frac{1}{\lambda} \cdot R^T \cdot u \right) = u.$$

Hence, the range of R contains at least $r + 1$ linearly independent vectors u and u_i, $i = 1, \dots, r$. This contradicts the fact that the rank of R is just r. Overall, we have shown that $R \cdot R^T$ has exactly r positive eigenvalues $\lambda_i = \sigma_i$, $i = 1, \dots, r$, and its rank is r.

It remains to show that $R^T \cdot R$ has the same eigenvalues λ_i, $i = 1, \ldots, r$, as $R \cdot R^T$. This holds true, since $v_i = R^T \cdot u_i$ is an eigenvector of $R^T \cdot R$ for the eigenvalue λ_i:

$$R^T \cdot R \cdot v_i = R^T \cdot \underbrace{R \cdot R^T \cdot u_i}_{= \lambda_i \cdot u_i} = \lambda_i \cdot \underbrace{R^T \cdot u_i}_{= v_i} = \lambda_i \cdot v_i.$$

Hence, both matrices share the same set of eigenvalues and the corresponding eigenvectors of $R \cdot R^T$ are the columns of U, while the eigenvectors of $R^T \cdot R$ make up the rows of V.

Exercise 3.3 (Frobenius Norm and Singular Values)
A straightforward calculation provides the assertion by using the singular value decomposition of $R = U \cdot \Sigma \cdot V$:

$$\|R\|_F^2 = \|U \cdot \Sigma \cdot V\|_F^2 = \text{trace}\left((U \cdot \Sigma \cdot V)^T \cdot (U \cdot \Sigma \cdot V)\right)$$

$$= \text{trace}\left(V^T \cdot \Sigma^T \cdot \underbrace{U^T \cdot U}_{=I} \cdot \Sigma \cdot V\right)$$

$$= \text{trace}\left(V^T \cdot \Sigma^T \cdot \Sigma \cdot V\right) = \text{trace}\left((\Sigma \cdot V)^T \cdot \Sigma \cdot V\right) = \text{trace}\left(\Sigma \cdot V \cdot (\Sigma \cdot V)^T\right)$$

$$= \text{trace}\left(\Sigma \cdot \underbrace{V \cdot V^T}_{=I} \cdot \Sigma^T\right) = \text{trace}\left(\Sigma^2\right) = \sigma_1^2 + \sigma_2^2 + \ldots + \sigma_r^2.$$

Exercise 3.4 (Largest and Smallest Singular Values)
We solve an equivalent optimization problem:

$$\max_z \ \|R \cdot z\|_2^2 \quad \text{s.t.} \quad \|z\|^2 - 1 = 0.$$

For that, the Lagrange multiplier rule is applied, see e.g. Jongen et al. (2004):

$$\nabla \|R \cdot z\|_2^2 = \lambda \cdot \nabla \left(\|z\|^2 - 1\right).$$

By a straightforward computation, we obtain:

$$R^T \cdot R \cdot z = \lambda \cdot z.$$

We see that z is an eigenvector of the matrix $R^T \cdot R$ corresponding to the eigenvalue λ. We further deduce:

$$\|R \cdot z\|_2^2 = z^T \cdot \underbrace{R^T \cdot R \cdot z}_{= \lambda \cdot z} = \lambda \cdot \underbrace{\|z\|_2^2}_{=1} = \lambda.$$

This means that λ must be the largest eigenvalue of the positive semidefinite matrix $R^T \cdot R$, i.e.

$$\lambda = \lambda_{max}\left(R^T \cdot R\right).$$

Hence, we get:

$$\|R \cdot z\|_2 = \sqrt{\lambda_{max}\left(R^T \cdot R\right)}.$$

In view of Exercise 3.2, the latter is the largest singular value of the matrix R, i.e.

$$\sqrt{\lambda_{max}\left(R^T \cdot R\right)} = \sigma_{max}(R).$$

Altogether, we have shown the assertion:

$$\max_{\|z\|_2=1} \|R \cdot z\|_2 = \sigma_{max}(R).$$

The formula for the smallest singular value of R follows analogously.

Exercise 3.5 (Features)
(i) Recall the user-movie rating matrix

$$R = \begin{pmatrix} 1 & 2 & 4 & -9 \\ 1 & 1 & 3 & -6 \\ -5 & 6 & -4 & -3 \end{pmatrix}.$$

In order to find the number of latent features, we calculate its rank by Gaussian elimination:

$$\begin{matrix} 1 & 2 & 4 & -9 \\ 1 & 1 & 3 & -6 \\ -5 & 6 & -4 & -3 \end{matrix} \sim \begin{matrix} 1 & 2 & 4 & -9 \\ 0 & -1 & -1 & 3 \\ 0 & 16 & 16 & -48 \end{matrix} \sim \begin{matrix} 1 & 2 & 4 & -9 \\ 0 & -1 & -1 & 3 \\ 0 & 0 & 0 & 0 \end{matrix}$$

We conclude that $\text{rank}(R) = 2$ and thus there are two latent features. The importance of latent features can be read from the singular values of R. The singular values of R are the square roots of the eigenvalues of $R \cdot R^T$, see Exercise 3.5:

$$R \cdot R^T = \begin{pmatrix} 102 & 69 & 18 \\ 69 & 47 & 7 \\ 18 & 7 & 86 \end{pmatrix}.$$

The characteristic polynomial of $R \cdot R^T$ is

$$\det \left(R \cdot R^T - \lambda \cdot I \right) = -\lambda^3 + 235 \cdot \lambda^2 - 12{,}474 \cdot \lambda.$$

Its non-zero roots are $\lambda_1 = 154$ and $\lambda_2 = 81$, hence, the singular values of R are

$$\sigma_1 = \sqrt{154}, \quad \sigma_2 = 9.$$

(ii) The singular value decomposition $R = U \cdot \Sigma \cdot V$ can be computed by hand. Note that the columns of U are the eigenvectors of $R \cdot R^T$, and the rows of V are the eigenvectors of $R^T \cdot R$ corresponding to the eigenvalues $\lambda_1 = \sigma_1^2$ and $\lambda_2 = \sigma_2^2$, see Exercise 3.5. Singular values are stored in the diagonal matrix $\Sigma = \mathrm{diag}\,(\sigma_1, \sigma_2)$. By using one's favorite software, we obtain:

$$\underbrace{\begin{pmatrix} 1\ 2 & 4 & -9 \\ 1\ 1 & 3 & -6 \\ -5\ 6 & -4 & -3 \end{pmatrix}}_{R} = \underbrace{\begin{pmatrix} \frac{3}{\sqrt{14}} & -\frac{1}{\sqrt{27}} \\ \frac{2}{\sqrt{14}} & -\frac{1}{\sqrt{27}} \\ \frac{1}{\sqrt{14}} & \frac{5}{\sqrt{27}} \end{pmatrix}}_{U} \cdot \underbrace{\begin{pmatrix} \sqrt{154} & 0 \\ 0 & 9 \end{pmatrix}}_{\Sigma} \cdot \underbrace{\begin{pmatrix} 0 & \frac{1}{\sqrt{11}} & \frac{1}{\sqrt{11}} & -\frac{3}{\sqrt{11}} \\ -\frac{1}{\sqrt{3}} & \frac{1}{\sqrt{3}} & -\frac{1}{\sqrt{3}} & 0 \end{pmatrix}}_{V}.$$

Due to Eckart-Young-Mirsky theorem, a best 1-rank approximation of R is given by

$$A = U \cdot \Sigma_1 \cdot V,$$

where in $\Sigma_1 = \mathrm{diag}\,(\sigma_1, 0)$ the smallest singular value is set to zero. Hence, we obtain:

$$A = \underbrace{\begin{pmatrix} \frac{3}{\sqrt{14}} & -\frac{1}{\sqrt{27}} \\ \frac{2}{\sqrt{14}} & -\frac{1}{\sqrt{27}} \\ \frac{1}{\sqrt{14}} & \frac{5}{\sqrt{27}} \end{pmatrix}}_{U} \cdot \underbrace{\begin{pmatrix} \sqrt{154} & 0 \\ 0 & 0 \end{pmatrix}}_{\Sigma} \cdot \underbrace{\begin{pmatrix} 0 & \frac{1}{\sqrt{11}} & \frac{1}{\sqrt{11}} & -\frac{3}{\sqrt{11}} \\ -\frac{1}{\sqrt{3}} & \frac{1}{\sqrt{3}} & -\frac{1}{\sqrt{3}} & 0 \end{pmatrix}}_{V}.$$

The corresponding approximation error is the smallest singular value of R:

$$\| R - A \|_F = \sqrt{\sigma_2^2} = \sigma_2 = 9.$$

Exercise 3.6 (Rank of Matrix Product)
Since the rank of Y is s, there are s linear independent vectors within the range of Y:

$$y_i = Y \cdot z_i, \quad i = 1, \ldots s.$$

We set:

$$x_i = X \cdot y_i, \quad i = 1, \dots, s.$$

Note that these vectors are within the range of $X \cdot Y$:

$$x_i = X \cdot y_i = X \cdot Y \cdot z_i, \quad i = 1, \dots s.$$

Let us show that they are linearly independent. If not, there exist real numbers $c_1, \dots, c_r$ not all vanishing simultaneously, such that it holds:

$$\sum_{i=1}^{s} c_i \cdot x_i = 0.$$

By substituting, we proceed:

$$\sum_{i=1}^{s} c_i \cdot X \cdot y_i = X \cdot \left(\sum_{i=1}^{s} c_i \cdot y_i \right) = 0.$$

Since all s columns of X are linearly independent, we obtain from here:

$$\sum_{i=1}^{s} c_i \cdot y_i = 0.$$

Since y_i's are linearly independent by construction, we have $c_1 = \dots = c_r = 0$, a contradiction. Overall, the vectors $x_1, \dots, x_r$ from the range of $X \cdot Y$ are linearly independent. If there are more than s linear independent vectors within the range of $X \cdot Y$, then they also are within the range of X, a contradiction to $\dim(\text{range}(X)) = s$. Hence, we have:

$$\text{rank}(X \cdot Y) = \dim(\text{range}(X \cdot Y)) = s.$$

Exercise 3.7 (Low-Rank Approximation)
We sketch the steps of $(\mathcal{GD})$ and refer to the Python code for the numerical results. We note that the low-rank approximation problem is not convex and therefore global convergence of the gradient descent is not guaranteed. Hence, the application of $(\mathcal{GD})$

is very sensitive with respect to initialization and the choice of a step size. Let us present the implementation for the step size $\eta = 0.1$ and starting matrices

$$
X(0) = \begin{pmatrix} 0 & 0.3 \\ 0 & 0 \\ 0 & 0.1 \\ 0 & 0 \\ 0.1 & 0 \end{pmatrix}, \quad Y(0) = \begin{pmatrix} 0 & 0 & 0.1 & 0 \\ 0 & 0.1 & 0 & 0.1 \end{pmatrix}.
$$

Note that both matrices have rank equal to 2. In order to calculate the matrices $X(1)$ and $Y(1)$, it remains to determine:

$$
E(0) = \begin{pmatrix} 5 & 2.97 & 0 & 0.97 \\ 4 & 0 & 0 & 1 \\ 1 & 0.99 & 0 & 4.99 \\ 1 & 0 & 0 & 4 \\ 1 & 1 & 4.99 & 4 \end{pmatrix}.
$$

Then we have:

$$
X(1) = X(0) + \eta \cdot E(0) \cdot Y^T(0) = \begin{pmatrix} 0 & 0.4267 \\ 0 & 0.01 \\ 0 & 0.1598 \\ 0 & 0.04 \\ 0.1499 & 0.05 \end{pmatrix},
$$

$$
Y(1) = Y(0) + \eta \cdot X^T(0) \cdot E(0) = \begin{pmatrix} 0.01 & 0.01 & 0.1499 & 0.04 \\ 0.16 & 0.199 & 0 & 0.179 \end{pmatrix}.
$$

After 15 iterations, $(\mathcal{GD})$ yields:

$$
R_{model} = X(15) \cdot Y(15) =
$$

	M1	M2	M3	M4
U1	4.99	2.97	**1.25**	0.94
U2	3.97	**2.38**	**1.24**	0.96
U3	0.96	1.01	**5.54**	4.81
U4	0.99	**0.93**	**4.44**	3.85
U5	**0.97**	0.95	**4.76**	4.13

We note that there are several variants for this completion, e.g. the objective function can be regularized to avoid overfitting. Besides, other algorithms such as alternating least squares or stochastic gradient descent can be applied.

10.4 Classification

Exercise 4.1 (Fisher's Discriminant)
Let the clients be indexed by $i = 1, \ldots, 6$. Due to the creditworthiness, we divide them into the classes:

$$C_{yes} = \{1, 2, 3, 5\}, \quad C_{no} = \{4, 6\}.$$

Clearly, we have $n = 6$, $n_{yes} = 4$, and $n_{no} = 2$. Next, we calculate the sample means:

$$\bar{x} = \frac{1}{6} \cdot \sum_{i=1}^{n} x_i = \begin{pmatrix} 4 \\ 20 \\ 27/3 \end{pmatrix},$$

$$x_{yes} = \frac{1}{4} \cdot \sum_{i \in C_{yes}} x_i = \begin{pmatrix} 5 \\ 12.5 \\ 2.5 \end{pmatrix}, \quad x_{no} = \frac{1}{2} \cdot \sum_{i \in C_{no}} x_i = \begin{pmatrix} 2 \\ 35 \\ 2 \end{pmatrix}.$$

Note that the feature space is 3-dimensional, i.e. $x_i \in \mathbb{R}^3$, $i = 1, \ldots, 6$, hence, the matrix W is of order 3×3. Recalling the formula

$$W = \frac{1}{n} \cdot \left(\sum_{i \in C_{yes}} (x_i - x_{yes}) \cdot (x_i - x_{yes})^T + \sum_{i \in C_{no}} (x_i - x_{no}) \cdot (x_i - x_{no})^T \right),$$

we get the matrix (here and later on we round up to four digits):

$$W = \begin{pmatrix} 2.6667 & -5.0000 & -1.1667 \\ -5.0000 & 54.1667 & 5.8333 \\ -1.1667 & 5.8333 & 0.8333 \end{pmatrix}.$$

In order to calculate the Fisher's discriminant, we exploit the derived formula for the eigenvector a corresponding to the largest eigenvalue of $W^{-1} \cdot B$, which reads:

$$a = W^{-1} \cdot (x_{yes} - x_{no}).$$

Thus, we conclude that the Fisher discriminant is

$$a = \begin{pmatrix} 35.1111 \\ -10.2889 \\ 121.7778 \end{pmatrix}.$$

It remains to determine the bound:

$$b = \frac{\left(x_{yes} - x_{no}\right)^T \cdot W^{-1} \cdot \left(x_{yes} + x_{no}\right)}{2}.$$

Plugging the values into the formula provides:

$$b = 152.5278.$$

The new client can be represented as a 3-dimensional feature vector, i.e.

$$x = \begin{pmatrix} 4 \\ 10 \\ 1 \end{pmatrix}.$$

We conclude that the newcomer x is labeled by $y = 1$, since

$$a^T \cdot x = 159.3332 \geq 152.5278 = b.$$

In other words, the client should be considered as creditworthy.

Exercise 4.2 (Sample Means)
Let us start by showing that the first term vanishes, i.e.

$$\sum_{i \in C_{yes}} \left(z_i - z_{yes}\right) \cdot \left(z_{yes} - \bar{z}\right) = 0.$$

Recalling the formula for the sample mean

$$z_{yes} = \frac{1}{n_{yes}} \cdot \sum_{i \in C_{yes}} z_i,$$

a straightforward calculation provides:

$$\sum_{i\in C_{yes}} (z_i - z_{yes}) \cdot (z_{yes} - \bar{z}) = \sum_{i\in C_{yes}} z_i \cdot (z_{yes} - \bar{z}) - \sum_{i\in C_{yes}} z_{yes} \cdot (z_{yes} - \bar{z})$$

$$= (z_{yes} - \bar{z}) \cdot n_{yes} \cdot z_{yes} - (z_{yes} - \bar{z}) \cdot n_{yes} \cdot z_{yes} = 0.$$

The same can be derived for the second term:

$$\sum_{i\in C_{no}} (z_i - z_{no}) \cdot (z_{no} - \bar{z}) = 0.$$

Exercise 4.3 (Homogeneous Functions)
We show first that

$$\max_a \frac{f(a)}{g(a)} = \max_a \ \{f(a) \,|\, g(a) = 1\}.$$

Let a be a solution of the left optimization problem. Then, $t \cdot a$ does also solve it for all $t > 0$. This follows due to the homogeneity of f and g of degree α:

$$\frac{f(t \cdot a)}{g(t \cdot a)} = \frac{t^\alpha \cdot f(a)}{t^\alpha \cdot g(a)} = \frac{f(a)}{g(a)}.$$

Since $g(a) > 0$, we may set:

$$t = \sqrt[\alpha]{\frac{1}{g(a)}}.$$

Then, $t \cdot a$ is feasible for the right optimization problem:

$$g(t \cdot a) = t^\alpha \cdot g(a) = \left(\sqrt[\alpha]{\frac{1}{g(a)}}\right)^\alpha \cdot g(a) = 1.$$

Moreover, for its objective function we have:

$$f(t \cdot a) = t^\alpha \cdot f(a) = \left(\sqrt[\alpha]{\frac{1}{g(a)}}\right)^\alpha \cdot f(a) = \frac{f(a)}{g(a)}.$$

This proves:

$$\max_a \frac{f(a)}{g(a)} \leq \max_a \ \{f(a) \,|\, g(a) = 1\}.$$

The reverse inequality is trivial.

Now, we turn our attention to the optimization problems $(\mathcal{V})$ and $(\mathcal{E})$. Note that the optimization problem $(\mathcal{V})$ is of the form

$$\max_{a} \ \frac{B(a)}{W(a)},$$

where

$$B(a) = a^T \cdot B \cdot a, \quad W(a) = a^T \cdot W \cdot a.$$

Furthermore, the function $B(a)$ is homogeneous of degree two, since for $t \geq 0$ it holds:

$$B(t \cdot a) = (t \cdot a)^T \cdot B \cdot (t \cdot a) = t^2 \cdot a^T \cdot B \cdot a = t^2 \cdot B(a).$$

Clearly, the same is also true for the function $W(a)$. Besides, the positive semidefinite matrix W is regular by assumption and, thus, positive definite, i.e. $W(a) > 0$ for all $a \in \mathbb{R}^m$. Hence, the above result applies. It says that $(\mathcal{V})$ is equivalent to the optimization problem $(\mathcal{E})$:

$$\max_{a} \ a^T \cdot B \cdot a \quad \text{s.t.} \quad a^T \cdot W \cdot a = 1.$$

Exercise 4.4 (Hyperplanes)
Given two parallel hyperplanes $a^T \cdot x - b_1 = 0$ and $a^T \cdot x - b_2 = 0$, we calculate the distance between them. For that, we take an arbitrary point x_1 on the first hyperplane. From this point we go along the direction of the normal vector a. As the hyperplanes are parallel, moving along the direction a results in an intersection with the second hyperplane at a point x_2, i.e. for some $t \in \mathbb{R}$ we have:

$$x_2 = x_1 + t \cdot a.$$

It remains to determine how far we have to move along the direction of the normal vector, i.e. we calculate the value of t. Since x_2 lies on the second hyperplane, we get:

$$a^T \cdot (x_1 + t \cdot a) = b_2.$$

As the point x_1 lies on the first hyperplane:

$$\underbrace{a^T \cdot x_1}_{=b_1} + t \cdot \underbrace{a^T \cdot a}_{=\|a\|_2^2} = b_2.$$

Simplifying further provides:

$$t = \frac{b_2 - b_1}{\|a\|_2^2}.$$

We are ready to calculate the distance between the two hyperplanes:

$$\|x_2 - x_1\|_2 = \left\| x_1 + \frac{b_2 - b_1}{\|a\|_2^2} \cdot a - x_1 \right\|_2 = \frac{|b_2 - b_1| \cdot \|a\|_2}{\|a\|_2^2} = \frac{|b_2 - b_1|}{\|a\|_2}.$$

Recall that the margin is the half-width of the region bounded by the two parallel hyperplanes:

$$a^T \cdot x - b = +\gamma, \quad a^T \cdot x - b = -\gamma.$$

We define $b_1 = b + \gamma$ and $b_2 = b - \gamma$, which yields the distance:

$$\frac{|b_2 - b_1|}{\|a\|_2} = \frac{|b - \gamma - (b + \gamma)|}{\|a\|_2} = \frac{2 \cdot \gamma}{\|a\|_2}.$$

Hence, the half-width is $\frac{\gamma}{\|a\|_2}$, and the assertion holds.

Exercise 4.5 (Regularized SVM)
We dualize the problem ($\mathcal{P}_{reg}$). For that, let us introduce the Lagrange multipliers λ_i, $i = 1, \ldots, n$. We get, see e.g. Nesterov (2018):

$$\max_{\lambda \geq 0} \min_{a,b,\xi \geq 0} \frac{1}{2} \cdot \|a\|_2^2 + c \cdot \sum_{i=1}^{n} \xi_i + \sum_{i=1}^{n} \lambda_i \cdot \left(1 - \xi_i - y_i \cdot \left(a^T \cdot x_i - b\right)\right).$$

The necessary optimality condition for the inner minimization problem with respect to a and b yields:

$$a = \sum_{i=1}^{n} \lambda_i \cdot y_i \cdot x_i, \quad \sum_{i=1}^{n} \lambda_i \cdot y_i = 0.$$

Minimization with respect to ξ provides for all $i = 1, \ldots, n$:

$$\xi_i \geq 0, \quad c - \lambda_i \geq 0, \quad (c - \lambda_i) \cdot \xi_i = 0.$$

Plugging into the objective function gives:

$$\frac{1}{2} \cdot \sum_{i,j=1}^{n} \lambda_i \cdot \lambda_j \cdot y_i \cdot y_j \cdot x_i^T \cdot x_j + c \cdot \sum_{i=1}^{n} \xi_i$$

$$+ \sum_{i=1}^{n} \lambda_i - \underbrace{\sum_{i=1}^{n} \lambda_i \cdot \xi_i}_{=c \cdot \xi_i} + b \cdot \underbrace{\sum_{i=1}^{n} \lambda_i \cdot y_i}_{=0} - \sum_{i,j=1}^{n} \lambda_i \cdot \lambda_j \cdot y_i \cdot y_j \cdot x_i^T \cdot x_j.$$

We conclude that the dual problem $(\mathcal{D}_{reg})$ reads:

$$\max_{c \cdot e \geq \lambda \geq 0} \sum_{i=1}^{n} \lambda_i - \frac{1}{2} \cdot \sum_{i,j=1}^{n} \lambda_i \cdot \lambda_j \cdot y_i \cdot y_j \cdot x_i^T \cdot x_j \quad \text{s.t.} \quad \sum_{i=1}^{n} \lambda_i \cdot y_i = 0.$$

Exercise 4.6 (Kernel Rules)
We analyze the sum of two valid kernels

$$S(u, v) = K(u, v) + L(u, v).$$

Each of them can be written as a scalar product of feature mappings, i.e.

$$K(u, v) = \phi(u)^T \cdot \phi(v), \quad L(u, v) = \varphi(u)^T \cdot \varphi(v),$$

where $\phi : \mathbb{R}^m \mapsto \mathbb{R}^{r_K}$ and $\varphi : \mathbb{R}^m \mapsto \mathbb{R}^{r_L}$. Let us concatenate both feature mappings:

$$\Phi(u) = [\phi(u), \varphi(u)].$$

It follows that we can express S as a scalar product of $\Phi : \mathbb{R}^m \mapsto \mathbb{R}^{r_K + r_L}$:

$$S(u, v) = K(u, v) + L(u, v) = \phi(u)^T \cdot \phi(v) + \varphi(u)^T \cdot \varphi(v)$$

$$= [\phi(u), \varphi(u)]^T \cdot [\phi(v), \varphi(v)] = \Phi(u)^T \cdot \Phi(v).$$

Hence, S is valid with the intrinsic degree $r_S = r_K + r_L$.
    Next, we show that the product of two valid kernels is also valid. Denoting the feature mappings of K and L again as ϕ and φ, respectively, straightforward calculations provide:

$$P(u, v) = K(u, v) \cdot L(u, v) = \left(\phi(u)^T \cdot \phi(v)\right) \cdot \left(\varphi(u)^T \cdot \varphi(v)\right)$$

$$= \sum_{k=1}^{r_K} \phi_k(u) \cdot \phi_k(v) \cdot \sum_{\ell=1}^{r_L} \varphi_\ell(u) \cdot \varphi_\ell(v) = \sum_{k=1}^{r_K} \sum_{\ell=1}^{r_\ell} \phi_k(u) \cdot \varphi_\ell(u) \cdot \phi_k(v) \cdot \varphi_\ell(v).$$

We define $\Phi_{k\ell}(u) = \phi_k(u) \cdot \varphi_\ell(u)$ for all elements in the sum, which implies:

$$P(u, v) = \Phi(u)^T \cdot \Phi(v).$$

Hence, the kernel P is valid with the intrinsic degree $r_P = r_K \cdot r_L$.

Exercise 4.7 (Polynomial Kernel)
The polynomial kernel reads as

$$\left(u^T \cdot v\right)^d = \left(\sum_{i=1}^{m} u_i \cdot v_i\right)^d.$$

In order to derive the feature mapping, we expand this expression into the powers of terms in the sum. for that, we apply the multinomial theorem, which is a generalization of the binomial theorem. This yields:

$$\left(u^T \cdot v\right)^d = \sum_{k_1+k_2+\ldots+k_m=d} \frac{d!}{k_1! \cdots k_m!} \cdot \prod_{i=1}^{m} (u_i \cdot v_i)^{k_i},$$

where $k_1, \ldots, k_m$ are nonnegative integers and the sum goes over all their combinations yielding d. The terms $\frac{d!}{k_1! \cdots k_m!}$ are called *multinomial coefficients*. We further obtain:

$$\left(u^T \cdot v\right)^d = \sum_{k_1+k_2+\ldots+k_m=d} \frac{d!}{k_1! \cdots k_m!} \cdot \prod_{i=1}^{m} (u_i)^{k_i} \cdot \prod_{i=1}^{m} (v_i)^{k_i}.$$

Let us denote a single combination of nonnegative integers by $k = (k_1, \ldots, k_m)$. Hence, we are able to define the elements of a feature mapping by

$$\Phi_k(u) = \sqrt{\frac{d!}{k_1! \cdots \cdots k_m!}} \cdot \prod_{i=1}^{m} (u_i)^{k_i}.$$

Thus, the kernel simplifies to

$$\left(u^T \cdot v\right)^d = \sum_{e^T \cdot k = d} \Phi_k(u)^T \cdot \Phi_k(v).$$

Altogether, we have shown that the polynomial kernel can be written as the scalar product of a feature mapping and is therefore valid. It remains to determine the intrinsic degree. We need to count nonnegative integers $k_1, \ldots, k_m$ with $k_1+k_2+\ldots+k_m = d$. This corresponds to the number of arrangements of d unlabeled balls into m different boxes. The latter can be computed by applying the so-called *stars and bars* method from combinatorics. For that, let us represent d balls by stars and place them on a line, where the stars for the first bin will be taken from the left, followed by the stars for the second bin, and so on. Thus, an arrangement will be determined once it is known which is the first star going to the second bin, and the first star going to the third bin, and so on. This can be indicated by placing $m - 1$ separating bars between the stars:

$$\underbrace{\star\star\star}_{k_1=3}|\underbrace{}_{k_2=0}|\underbrace{\star\star}_{k_3=2}|\star\star\star\ldots\star.$$

Observe that the desired arrangements consist of $d + m - 1$ objects. To fix a concrete arrangement, we just need to substitute $m - 1$ objects by bars. The number of those substitutions is

$$\binom{d + m - 1}{d}.$$

Overall, the intrinsic degree of the polynomial kernel is

$$r = \binom{d + m - 1}{m - 1}.$$

We note that the quadratic kernel with $d = 2$ is a special case with the number of features:

$$r = \binom{2 + m - 1}{2} = \binom{m + 1}{2}.$$

10.5 Clustering

Exercise 5.1 (k-Means Clustering)
Given the data points

$$x_1 = (1, 0)^T, \quad x_2 = (2, 0)^T, \quad x_3 = (3, 0)^T, \quad x_4 = (4, 0)^T, \quad x_5 = (5, 0)^T, \quad x_6 = (5, 1)^T$$

and centers $z_1 = (3, 0)^T$ and $z_2 = (5, 1)^T$ of the clusters, we start with step (1) in k-means. Since $k = 2$, we assign a data point x_i to the cluster C_1 if $\|x_i - z_1\|_2 < \|x_i - z_2\|_2$, and to the cluster C_2 otherwise.

$t = 1$

	$\|x_i - z_1\|_2$	$\|x_i - z_2\|_2$	Cluster
$i = 1$	2	$\sqrt{17}$	C_1
$i = 2$	1	$\sqrt{10}$	C_1
$i = 3$	0	$\sqrt{5}$	C_1
$i = 4$	1	$\sqrt{2}$	C_1
$i = 5$	2	1	C_2
$i = 6$	$\sqrt{5}$	0	C_2

Note that $|C_1| = 4$ and $|C_2| = 2$. In Step (2), we calculate the new centers:

$$z_1 = \frac{1}{4} \cdot (x_1 + x_2 + x_3 + x_4) = \frac{1}{4} \cdot \begin{pmatrix} 1+2+3+4 \\ 0+0+0+0 \end{pmatrix} = \begin{pmatrix} 2.5 \\ 0 \end{pmatrix},$$

$$z_2 = \frac{1}{2} \cdot (x_5 + x_6) = \frac{1}{2} \cdot \begin{pmatrix} 5+5 \\ 0+1 \end{pmatrix} = \begin{pmatrix} 5 \\ 0.5 \end{pmatrix}.$$

We repeat by using the new centers $z_1 = (2.5, 0)^T$ and $z_2 = (5, 0.5)$.

$t = 2$

	$\|x_i - z_1\|_2$	$\|x_i - z_2\|_2$	Cluster
$i = 1$	1.5	$\sqrt{16.25}$	C_1
$i = 2$	0.5	$\sqrt{9.25}$	C_1
$i = 3$	0.5	$\sqrt{4.25}$	C_1
$i = 4$	1.5	$\sqrt{1.25}$	C_2
$i = 5$	2.5	0.5	C_2
$i = 6$	$\sqrt{7.25}$	0.5	C_2

Step (2) provides:

$$z_1 = \frac{1}{3} \cdot (x_1 + x_2 + x_3) = \frac{1}{3} \cdot \begin{pmatrix} 1+2+3 \\ 0+0+0 \end{pmatrix} = \begin{pmatrix} 2 \\ 0 \end{pmatrix},$$

$$z_2 = \frac{1}{3} \cdot (x_4 + x_5 + x_6) = \frac{1}{3} \cdot \begin{pmatrix} 4+5+5 \\ 0+0+1 \end{pmatrix} = \begin{pmatrix} 14/3 \\ 1/3 \end{pmatrix}.$$

After updating the centers by $z_1 = (2, 0)^T$ and $z_2 = (14/3, 1/3)^T$, we go to the next iteration.

$t = 3$

	$\|x_i - z_1\|_2$	$\|x_i - z_2\|_2$	Cluster
$i = 1$	1	$\frac{\sqrt{122}}{3}$	C_1
$i = 2$	0	$\frac{\sqrt{65}}{3}$	C_1
$i = 3$	1	$\frac{\sqrt{26}}{3}$	C_1
$i = 4$	2	$\frac{\sqrt{5}}{3}$	C_2
$i = 5$	3	$\frac{\sqrt{2}}{3}$	C_2
$i = 6$	$\sqrt{10}$	$\frac{\sqrt{5}}{3}$	C_2

The data points are assigned to the same clusters as in the iteration before. Hence, the algorithm stops here, since the update step (2) would provide the same centers as calculated in iteration $t = 2$. This assignment of cluster and centers indeed solves the optimization problem $\mathcal{D}$.

Exercise 5.2 (Marginal Median)
Recall the definition of a median. We assume that N ordered numbers $a_1 < \ldots < a_N$ are given. Their median is defined as follows:

$$a = \begin{cases} a_{M+1}, & \text{if } N = 2M + 1, \\[2ex] \dfrac{a_M + a_{M+1}}{2}, & \text{if } N = 2M. \end{cases}$$

Let us turn our attention to the update step (2) in k-means:

$$\min_z \; d\left(C_\ell, z\right),$$

where the objective function is

$$d\left(C_\ell, z\right) = \sum_{i \in C_\ell} \|z - x_i\|_1 = \sum_{i \in C_\ell} \sum_{j=1}^m \left|z_j - (x_i)_j\right| = \sum_{j=1}^m \sum_{i \in C_\ell} \left|z_j - (x_i)_j\right|.$$

The convex subdifferential of the absolute value function, see e.g. Rockafellar (1970), is

$$\partial |y| = \begin{cases} 1, & \text{if } y > 0, \\ [-1, 1], & \text{if } y = 0, \\ -1, & \text{if } y < 0. \end{cases}$$

Hence, the j-th component of the convex subdifferential $\partial d\,(C_\ell, z)$ is given by

$$\frac{\partial d\,(C_\ell, z)}{\partial z_j} = \sum_{i \in C_\ell} \partial \left| z_j - (x_i)_j \right|.$$

In particular, we have by using the sign function:

$$\sum_{i \in C_\ell} \text{sign}\left(z_j - (x_i)_j \right) \in \frac{\partial d\,(C_\ell, z)}{\partial z_j}.$$

We set $(z_\ell)_j$ to be the median of the numbers $(x_i)_j$, $i \in C_\ell$. Thus, for all $j = 1, \ldots, m$ it holds:

$$\sum_{i \in C_\ell} \text{sign}\left((z_\ell)_j - (x_i)_j \right) = \#\left\{ i : (z_\ell)_j > (x_i)_j \right\} - \#\left\{ i : (z_\ell)_j < (x_i)_j \right\} = 0.$$

Overall, the necessary and sufficient optimality condition is satisfied:

$$0 \in \partial d\,(C_\ell, z),$$

where $z_\ell = ((z_\ell)_1, \ldots, (z_\ell)_m)^T$ is the cluster's C_ℓ marginal median.

Exercise 5.3 (Eigenvalues of a Stochastic Matrix)
Let the $(n \times n)$-matrix P be stochastic. A straightforward calculation provides that P and P^T have the same characteristic polynomial:

$$\det\left(P^T - I \cdot \lambda \right) = \det\,(P - I \cdot \lambda)^T = \det\,(P - I \cdot \lambda).$$

Hence, both matrices have the same eigenvalues. Let us show that the eigenvalues of P^T are bounded. Clearly, 1 is one of its eigenvalues:

$$P^T \cdot e = \left(e^T \cdot P \right)^T = \left(e^T \right)^T = 1 \cdot e.$$

Suppose that there exists an eigenvalue λ with $|\lambda| > 1$ and the corresponding eigenvector $v \neq 0$. We denote the maximal absolute value of its entries by

$$v_{\max} = \max_{j=1,\ldots,n} \left| v_j \right|.$$

Since P is a stochastic matrix, it holds for all $j = 1, \ldots, n$:

$$\left| \left(P^T \cdot v \right)_j \right| = \left| \sum_{i=1}^{n} p_{ij} \cdot v_i \right| \leq \sum_{i=1}^{n} p_{ij} \cdot \underbrace{|v_i|}_{\leq v_{max}} \leq \underbrace{\sum_{i=1}^{n} p_{ij}}_{=1} \cdot v_{max} = v_{max}.$$

On the other hand, we have for j with $|v_j| = v_{max} \neq 0$:

$$\left| \left(P^T \cdot v \right)_j \right| = \left| (\lambda \cdot v)_j \right| = \underbrace{|\lambda|}_{>1} \cdot |v_j| > |v_j| = v_{max}.$$

By comparing these two formulas we obtain a contradiction. Hence, the eigenvalues of P^T and, therefore, of P are bounded above by one.

Exercise 5.4 (Spectral Clustering)
We first calculate the matrix of transition probabilities:

$$P = W \cdot D^{-1} = \begin{pmatrix} 0 & 0.1111 & 0.3125 & 0 & 0.4 \\ 0.0625 & 0 & 0.5 & 0 & 0 \\ 0.3125 & 0.8889 & 0 & 0 & 0.12 \\ 0 & 0 & 0 & 0 & 0.48 \\ 0.625 & 0 & 0.1875 & 1 & 0 \end{pmatrix}.$$

Recall that D is a diagonal matrix with diagonal elements being the respective column sums of W. In order to determine the diffusion maps, we need to calculate $D^{-\frac{1}{2}}$. The latter matrix is given by

$$D^{-\frac{1}{2}} = \begin{pmatrix} 0.25 & 0 & 0 & 0 & 0 \\ 0 & 0.3333 & 0 & 0 & 0 \\ 0 & 0 & 0.25 & 0 & 0 \\ 0 & 0 & 0 & 0.28867513 & 0 \\ 0 & 0 & 0 & 0 & 0.2 \end{pmatrix}.$$

We diagonalize the matrix

$$S = V \cdot \Lambda \cdot V^T,$$

where

$$\Lambda = \mathrm{diag}\,(1, -0.8558, -0.7066, 0.6077 - 0.0454)$$

and

$$V = \begin{pmatrix} 0.4529 & -0.4137 & 0.2144 & 0.0021 & 0.76011 \\ 0.3397 & 0.0485 & 0.6558 & -0.5684 & -0.3594 \\ 0.4529 & -0.0106 & -0.7218 & -0.5184 & -0.0706 \\ 0.3922 & -0.57199 & -0.0381 & 0.4803 & -0.5356 \\ 0.5661 & 0.7066 & 0.0388 & 0.4213 & 0.0351 \end{pmatrix}.$$

Next, we are able to calculate the matrices Φ and Ψ as

$$\Phi = D^{\frac{1}{2}} \cdot V = \begin{pmatrix} 1.8116 & -1.6546 & 0.8577 & 0.0085 & 3.0405 \\ 1.019 & 0.1456 & 1.9673 & -1.7052 & -1.0782 \\ 1.8116 & -0.0425 & -2.8873 & -2.0737 & -0.2823 \\ 1.3587 & -1.9814 & -0.1319 & 1.6638 & -1.8553 \\ 2.8307 & 3.5329 & 0.1941 & 2.1066 & 0.1753 \end{pmatrix}$$

and

$$\Psi = D^{-\frac{1}{2}} \cdot V = \begin{pmatrix} 0.1132 & -0.1034 & 0.0536 & 0.0005 & 0.19 \\ 0.1132 & 0.0162 & 0.2186 & -0.1895 & -0.1198 \\ 0.1132 & -0.0027 & -0.1805 & -0.1296 & -0.0176 \\ 0.1132 & -0.1651 & -0.011 & 0.1386 & -0.1546 \\ 0.1132 & 0.1413 & 0.0078 & 0.0843 & 0.007 \end{pmatrix}.$$

Recall that the diffusion maps are defined as follows:

$$F_j(t) = \begin{pmatrix} \lambda_1^t \cdot (\psi_1)_j \\ \vdots \\ \lambda_5^t \cdot (\psi_5)_j \end{pmatrix}, \quad j = 1, \ldots, 5.$$

By setting $k = 2$, we get the k-truncated diffusion maps:

$$k\text{-}F_j(t) = \lambda_2^t \cdot (\psi_2)_j, \quad j = 1, \ldots, 5.$$

Let us now calculate the k-truncated diffusion maps after ten periods, i.e. for $t = 10$:

$$k\text{-}F_1(10) = \lambda_2^{10} \cdot (\psi_2)_1 = (-0.8558)^{10} \cdot (-0.1034) = -0.0218,$$

$$k\text{-}F_2(10) = \lambda_2^{10} \cdot (\psi_2)_2 = (-0.8558)^{10} \cdot 0.0162 = 0.0034,$$

$$k\text{-}F_3(10) = \lambda_2^{10} \cdot (\psi_2)_3 = (-0.8558)^{10} \cdot (-0.0027) = -0.0006,$$

$$k\text{-}F_4(10) = \lambda_2^{10} \cdot (\psi_2)_4 = (-0.8558)^{10} \cdot (-0.1651) = -0.0348,$$

$$k\text{-}F_4(10) = \lambda_2^{10} \cdot (\psi_2)_4 = (-0.8558)^{10} \cdot 0.1413 = 0.0298.$$

Spectral clustering consists of applying k-means in the Euclidean setup to the k-truncated diffusion maps. It remains to assign the numbers $k\text{-}F_j(10)$, $j = 1, \ldots, 5$ to the one cluster or another. We conclude that persons 1, 4 and persons 2, 3, 5 are clustered together. We refer to the Python code for numerical computations.

Exercise 5.5 (Time Series Clustering)
By using the Pearson correlation coefficient, the dissimilarity measure is

$$d(x, z) = \frac{1 - \text{Pearson}(x, z)}{2}.$$

The update step (2) in k-means reads:

$$z_\ell \in \arg\min_z \sum_{i \in C_\ell} \frac{1 - \text{Pearson}(x_i, z)}{2}.$$

This simplifies to

$$\max_{z \in \mathbb{R}^m} \sum_{i \in C_\ell} \frac{(x_i - \bar{x}_i)^T \cdot (z - \bar{z})}{\|x_i - \bar{x}_i\|_2 \cdot \|z - \bar{z}\|_2},$$

where $\bar{x}_i$ and $\bar{z}$ denote the means of the vectors x_i and z, respectively. Let us define

$$a = \sum_{i \in C_\ell} \frac{(x_i - \bar{x}_i)}{\|x_i - \bar{x}_i\|_2}.$$

Further we set $y = z - \bar{z}$ and note that its mean vanishes:

$$\bar{y} = \frac{1}{m} \cdot \sum_{j=1}^{m} y_j = \frac{1}{m} \cdot \sum_{j=1}^{m} (z_j - \bar{z}) = \bar{z} - \bar{z} = 0.$$

The optimization problem

$$\max_{z \in \mathbb{R}^m} a^T \cdot \frac{(z - \bar{z})}{\|z - \bar{z}\|_2}$$

is then equivalent to

$$\max_{\bar{y}=0} a^T \cdot \frac{y}{\|y\|_2} \quad \underset{w=\frac{y}{\|y\|_2}}{\Leftrightarrow} \quad \max_{\substack{\|w\|_2=1 \\ \bar{w}=0}} a^T \cdot w \quad \Leftrightarrow \quad \max_{\substack{\|w\|_2=1 \\ e^T \cdot w=0}} a^T \cdot w.$$

We solve the latter optimization problem by the Lagrange multiplier rule, see e.g. Jongen et al. (2004):

$$a = \lambda \cdot w + \mu \cdot e.$$

Multiplying by e from the left provides:

$$e^T \cdot a = \lambda \cdot \underbrace{e^T \cdot w}_{=0} + \mu \cdot \underbrace{e^T \cdot e}_{=m}.$$

Hence, we have:

$$\mu = \frac{1}{m} \cdot \sum_{j=1}^{m} a_j = \bar{a}.$$

Further, multiplying by w from the right yields:

$$a^T \cdot w = \lambda \cdot w^T \cdot w + \mu \cdot \underbrace{e^T \cdot w}_{=0}.$$

Thus, λ is the optimal value of the above optimization problem:

$$\lambda = a^T \cdot w.$$

Moreover, it follows:

$$|\lambda| = \|a - \mu \cdot e\|_2.$$

If $\lambda = 0$, we obtain $a = \mu \cdot e$, and any feasible w is optimal. Otherwise, $\lambda > 0$ due to the optimality of w, and we get:

$$w = \frac{a - \mu \cdot e}{\lambda} = \frac{a - \mu \cdot e}{\|a - \mu \cdot e\|_2} \cdot \underbrace{\text{sign}(\lambda)}_{=1} = \frac{a - \mu \cdot e}{\|a - \mu \cdot e\|_2} = \frac{a - \bar{a} \cdot e}{\|a - \bar{a} \cdot e\|_2}.$$

Let us turn our attention to $\bar{a}$. A straightforward calculation provides:

$$\bar{a} = \overline{\sum_{i \in C_\ell} \frac{(x_i - \bar{x}_i)}{\|x_i - \bar{x}_i\|_2}} = \sum_{i \in C_\ell} \overline{\left(\frac{(x_i - \bar{x}_i)}{\|x_i - \bar{x}_i\|_2}\right)} = \sum_{i \in C_\ell} \frac{1}{\|x_i - \bar{x}_i\|_2} \cdot \overline{(x_i - \bar{x}_i)} = 0.$$

Hence, the solution is given by

$$w = \frac{a}{\|a\|_2},$$

from where we conclude that the center of the cluster C_ℓ is

$$z_\ell = \sum_{i \in C_\ell} \frac{(x_i - \bar{x}_i)}{\|x_i - \bar{x}_i\|_2}.$$

We interpret the time series clustering as applied to stock prices. Each vector x_i then represents a time series of stock prices, i.e. the entry $(x_i)_j$ stands for the (log) returns of the i-th asset at time period j. The classes contain assets with mutually positive correlation. The latter can be used for diversifying a portfolio.

Exercise 5.6 (Product Clustering)

We face five products, which have been or have not been consumed by three customers. Hence, each product is represented as a binary vector $x_1, x_2, \ldots, x_5 \in \{0, 1\}^3$. The initial centers are $z_1 = (0, 1, 1)^T$ and $z_2 = (1, 1, 1)^T$. In order to measure the distance between two products represented as binary vectors x and y, we have to calculate the Jaccard coefficient $J(x, y)$. In our case, it is simply the share of customers $j \in \{1, 2, 3\}$ who bought both products compared to the customers who bought just one of them. Let us start with the Jaccard coefficient of the products with z_1:

$$J(x_1, z_1) = 1, \quad J(x_2, z_1) = \frac{1}{3}, \quad J(x_3, z_1) = \frac{1}{2}, \quad J(x_4, z_1) = 0, \quad J(x_5, z_1) = \frac{2}{3}.$$

From the latter we can directly calculate the corresponding dissimilarities:

$$d(x_1, z_1) = 0, \quad d(x_2, z_1) = \frac{2}{3}, \quad d(x_3, z_1) = \frac{1}{2}, \quad d(x_4, z_1) = 1, \quad d(x_5, z_1) = \frac{1}{3}.$$

The Jaccard coefficients of the products with z_2 are

$$J\left(x_1, z_2\right) = \frac{2}{3}, \quad J\left(x_2, z_2\right) = \frac{2}{3}, \quad J\left(x_3, z_2\right) = \frac{1}{3}, \quad J\left(x_4, z_2\right) = \frac{1}{3}, \quad J(x_5, z_2) = 1.$$

The corresponding dissimilarities are

$$d\left(x_1, z_2\right) = \frac{1}{3}, \quad d\left(x_2, z_2\right) = \frac{1}{3}, \quad d\left(x_3, z_2\right) = \frac{2}{3}, \quad d\left(x_4, z_2\right) = \frac{2}{3}, \quad d\left(x_5, z_2\right) = 0.$$

Hence, step (1) in k-means provides the following clusters:

$$C_1 = \{1, 3\}, \quad C_2 = \{2, 4, 5\}.$$

The centers also have to be binary vectors, therefore, step (2) is a combinatorial problem. We take all possible candidates and compute their distances to the newly defined clusters. The candidate which provides the smallest distance is the new center. For C_1 we get two possible centers $z_1 = (0, 0, 1)^T$ or $z_1 = (0, 1, 1)^T$. The new center for C_2 is given by $z_2 = (1, 0, 1)^T$. We proceed with $z_1 = (0, 0, 1)^T$ and $z_2 = (1, 0, 1)^T$ and leave the other option to the reader. Let us compute the Jaccard coefficient of the products with the new z_1:

$$J\left(x_1, z_1\right) = \frac{1}{2}, \quad J\left(x_2, z_1\right) = \frac{1}{2}, \quad J\left(x_3, z_1\right) = 1, \quad J\left(x_4, z_1\right) = 0, \quad J\left(x_5, z_1\right) = \frac{1}{3}.$$

The corresponding dissimilarities are

$$d\left(x_1, z_1\right) = \frac{1}{2}, \quad d\left(x_2, z_1\right) = \frac{1}{2}, \quad d\left(x_3, z_1\right) = 0, \quad d\left(x_4, z_1\right) = 1, \quad d\left(x_5, z_1\right) = \frac{2}{3}.$$

We calculate the Jaccard coefficients of the products with z_2:

$$J\left(x_1, z_2\right) = \frac{1}{3}, \quad J\left(x_2, z_2\right) = 1, \quad J\left(x_3, z_2\right) = \frac{1}{2}, \quad J\left(x_4, z_2\right) = \frac{1}{2}, \quad J(x_5, z_2) = \frac{2}{3}.$$

From where we compute the dissimilarities

$$d\left(x_1, z_2\right) = \frac{2}{3}, \quad d\left(x_2, z_2\right) = 0, \quad d\left(x_3, z_2\right) = \frac{1}{2}, \quad d\left(x_4, z_2\right) = \frac{1}{2}, \quad d\left(x_5, z_2\right) = \frac{1}{3}.$$

Hence the cluster assignment does not change in iteration 2, and we have finally:

$$C_1 = \{1, 3\}, \quad C_2 = \{2, 4, 5\}.$$

10.6   Linear Regression

Exercise 6.1 (Okun's Law)
Note that the Okun's law relates the changes of GDP to the changes of the unemployment rate. Therefore, in the first step we convert the panel data into the changes of the latter. We label each observation by index $i = 1, \ldots, 26$ and the corresponding GDP output and unemployment rate by GDP_i and u_i, respectively. In order to get the changes, we consider the differences:

$$\Delta_i GDP = GDP_{i+1} - GDP_i, \quad \Delta_i u = u_{i+1} - u_i \quad \text{for } i = 1, \ldots, 25.$$

Note that due to taking differences, the new data set reduces to 25 observations:

i	$\Delta_i GDP$	$\Delta_i u$
1	0.26	1.006
2	−0.05	1.352
3	0.14	1.053
4	0.38	−0.57
5	−0.09	0.667
6	−0.28	1.038
7	0.02	−0.075
8	−0.04	−0.933
9	−0.25	−0.938
10	0	−0.144
11	0.06	0.709
12	0.5	1.297
13	0.31	0.948
14	0.04	0.440
15	0.14	−0.917
16	0.44	−1.592
17	0.31	−1.134
18	−0.33	0.218
19	0	−0.776
20	0.34	−1.142
21	−0.22	−0.445
22	0.21	−0.148
23	0.13	−0.250
24	−0.52	−0.357
25	0.11	−0.502

We recall the statement of the Okun's law:

$$\Delta u = w_0 + w_1 \cdot \Delta\text{GDP} + \varepsilon,$$

where

$$\Delta\text{GDP} = (\Delta_i\text{GDP}, i = 1, \ldots, 25)^T, \quad \Delta u = (\Delta_i u, i = 1, \ldots, 25)^T.$$

Written in matrix form, we have:

$$\Delta u = X \cdot w + \varepsilon,$$

where X is a (25×2)-matrix with the first column consisting of the vector e of ones and the second column containing the entries of ΔGDP, i.e. $X = (e, \Delta\text{GDP})$. It remains to compute the OLS estimator

$$w_{OLS} = \left(X^T \cdot X\right)^{-1} \cdot X^T \cdot \Delta u = \begin{pmatrix} -0.0333 \\ -0.2259 \end{pmatrix}.$$

According to the panel data for Germany, the Okun's law can be confirmed with $w_1 = -0.2259$.

Exercise 6.2 (Pseudoinverse)
We recall the definition of the pseudoinverse $X^\dagger = \left(X^T \cdot X\right)^{-1} \cdot X^T$. The property (i) is valid, since

$$X \cdot X^\dagger \cdot X = X \cdot \underbrace{\left(X^T \cdot X\right)^{-1} \cdot X^T \cdot X}_{=I} = X.$$

Condition (ii) is fulfilled as

$$X^\dagger \cdot X \cdot X^\dagger = \underbrace{\left(X^T \cdot X\right)^{-1} \cdot X^T \cdot X}_{=I} \cdot \left(X^T \cdot X\right)^{-1} \cdot X^T = X^\dagger.$$

We turn our attention to (iii):

$$\left(X \cdot X^\dagger\right)^T = \left(X \cdot \left(X^T \cdot X\right)^{-1} \cdot X^T\right)^T = X \cdot \left(\left(X^T \cdot X\right)^{-1}\right)^T \cdot X^T$$

$$= X \cdot \left(\left(X^T \cdot X\right)^T\right)^{-1} \cdot X^T = X \cdot \left(X^T \cdot X\right)^{-1} \cdot X^T = X \cdot X^\dagger.$$

Exercise 6.3 (Hilbert Matrix)

Hilbert matrices are famous examples for ill-conditioned matrices. Let us present the comparison of the numerical and asymptotic results for their condition numbers:

$\kappa(H_n) = \frac{\sigma_{max}(H_n)}{\sigma_{min}(H_n)}$	$\frac{(1+\sqrt{2})^{4n}}{\sqrt{n}}$
H_5 : 476,607.2502422687	20,231,528.940628633
H_{10} : $1.60252853523e + 13$	$6.47183465942e + 14$
H_{15} : $1.10542932559e + 18$	$2.39053711440e + 22$

We notice that the condition number explodes with increasing n. In part (ii) of the exercise, we see that a high condition number leads to numerical instabilities, causing troubles to retain correct solutions. Note that the matrices H_5, H_{10}, H_{15} are regular, and the solution of the linear system

$$H_n \cdot x = H_n \cdot e, \quad n = 5, 10, 15,$$

is easy to determine as

$$x_n = H_n^{-1} \cdot H_n \cdot e = e, \quad n = 5, 10, 15.$$

Yet, the numerical computation is not able to provide the exact result, except for H_5. While for H_{10} the numerical calculation yields at least a vector close to e, in case of H_{15} the resulting vector varies from e a lot. This suggests that inverting an ill-conditioned matrix is numerically unstable, see the Python code for details.

Exercise 6.4 (Vandermonde Matrix)

We show by induction that for the determinant of the Vandermonde $(m \times m)$-matrix V it holds:

$$\det(V) = \prod_{1 \leq i < j \leq m} (\alpha_j - \alpha_i).$$

Clearly, the latter holds for $m = 1$, as the determinant of the Vandermonde (1×1)-matrix is 1, while the product on the right hand side is empty and, therefore, also equals to 1. For the base of induction, we can go even further and immediately see, that equality is fulfilled for the Vandermonde (2×2)-matrix. The product then yields $\alpha_2 - \alpha_1$, which is obviously the determinant of the Vandermonde matrix in this case. Let us try to compute the determinant by using the induction hypothesis that the formula holds for the Vandermonde $((m-1) \times (m-1))$-matrix. We compute the determinant for the dimension m:

$$\det(V) = \begin{vmatrix} 1 & \alpha_1 & \alpha_1^2 & \cdots & \alpha_1^{m-1} \\ 1 & \alpha_2 & \alpha_2^2 & \cdots & \alpha_2^{m-1} \\ \vdots & \vdots & \vdots & \ddots & \vdots \\ 1 & \alpha_m & \alpha_m^2 & \cdots & \alpha_m^{m-1} \end{vmatrix}.$$

We denote by A_i the i-th column of V. As row and column operations do not change the determinant, we take $A_i - \alpha_m \cdot A_{i-1}$ for each column, but the first. This yields:

$$\det(V) = \begin{vmatrix} 1 & \alpha_1 - \alpha_m & (\alpha_1 - \alpha_m) \cdot \alpha_1 & \cdots & (\alpha_1 - \alpha_m) \cdot \alpha_1^{m-2} \\ 1 & \alpha_2 - \alpha_m & (\alpha_2 - \alpha_m) \cdot \alpha_2 & \cdots & (\alpha_2 - \alpha_m) \cdot \alpha_2^{m-2} \\ \vdots & \vdots & \vdots & \ddots & \vdots \\ 1 & \alpha_{m-1} - \alpha_m & (\alpha_{m-1} - \alpha_m) \cdot \alpha_{m-1} & \cdots & (\alpha_{m-1} - \alpha_m) \cdot \alpha_{m-1}^{m-2} \\ 1 & 0 & 0 & \cdots & 0 \end{vmatrix}.$$

We expand the determinant with respect to the last row:

$$\det(V) = (-1)^{m+1} \cdot \begin{vmatrix} \alpha_1 - \alpha_m & (\alpha_1 - \alpha_m) \cdot \alpha_1 & \cdots & (\alpha_1 - \alpha_m) \cdot \alpha_1^{m-2} \\ \alpha_2 - \alpha_m & (\alpha_2 - \alpha_m) \cdot \alpha_2 & \cdots & (\alpha_2 - \alpha_m) \cdot \alpha_2^{m-2} \\ \vdots & \vdots & \ddots & \vdots \\ \alpha_{m-1} - \alpha_m & (\alpha_{m-1} - \alpha_m) \cdot \alpha_{m-1} & \cdots & (\alpha_{m-1} - \alpha_m) \cdot \alpha_{m-1}^{m-2} \end{vmatrix}.$$

We further apply determinant rules to get:

$$\det(V) = (-1)^{m+1} \cdot \prod_{i=1}^{m-1} (\alpha_i - \alpha_m) \cdot \begin{vmatrix} 1 & \alpha_1 & \alpha_1^2 & \cdots & \alpha_1^{m-2} \\ 1 & \alpha_2 & \alpha_2^2 & \cdots & \alpha_2^{m-2} \\ \vdots & \vdots & \vdots & \ddots & \vdots \\ 1 & \alpha_{m-1} & \alpha_{m-1}^2 & \cdots & \alpha_{m-1}^{m-2} \end{vmatrix}.$$

Note that the remaining determinant is known due to induction hypothesis:

$$\det(V) = (-1)^{m+1} \cdot \prod_{i=1}^{m-1} (\alpha_i - \alpha_m) \cdot \prod_{1 \le i < j \le m-1} (\alpha_j - \alpha_i).$$

In the next step, we change the order of the differences in the first product:

$$(-1)^{m+1} \cdot \prod_{i=1}^{m-1} (\alpha_i - \alpha_m) = (-1)^2 \cdot \prod_{i=1}^{m-1} (\alpha_m - \alpha_i) = \prod_{i=1}^{m-1} (\alpha_m - \alpha_i).$$

Therefore, we conclude for the determinant:

$$\det(V) = \prod_{i=1}^{m-1} (\alpha_m - \alpha_i) \cdot \prod_{1 \leq i < j \leq m-1} (\alpha_j - \alpha_i) = \prod_{1 \leq i < j \leq m} \left(\alpha_j - \alpha_i \right).$$

The assertion holds by induction principle.

Exercise 6.5 (Polynomial Regression)
It is important to note that the polynomial regression remains linear in the coefficients. Hence, we can treat each power of x as an exogenous variable by itself. Given a data set, the endogenous variable y_i depends on the exogenous variable x_i for $i = 1, \ldots, n$ in the following way:

$$y_i = w_0 + x_i \cdot w_1 + x_i^2 \cdot w_2 + \ldots + x_i^{m-1} \cdot w_{m-1} + \varepsilon_i.$$

The linear structure of this dependence enables us to write in matrix form:

$$y = X \cdot w + \varepsilon,$$

where $y \in \mathbb{R}^n$ is the data vector of endogenous variables and ε consists of n random errors:

$$y = (y_1, \ldots, y_n)^T, \quad \varepsilon = (\varepsilon_1, \ldots, \varepsilon_n)^T.$$

The vector of weights $w \in \mathbb{R}^m$ is given by

$$w = (w_0, w_1, \ldots, w_{m-1})^T.$$

Let us have a closer look on the data $(n \times m)$-matrix

$$X = \begin{pmatrix} 1 & x_1 & x_1^2 & \ldots & x_1^{m-1} \\ 1 & x_2 & x_2^2 & \ldots & x_2^{m-1} \\ \vdots & \vdots & \vdots & \ddots & \vdots \\ 1 & x_n & x_n^2 & \ldots & x_n^{m-1} \end{pmatrix}.$$

As $n > m$, we have to decide if $\operatorname{rank}(X) = m$, in order to ensure the lack of multicollinearity. We notice that the first m rows of X constitute a Vandermonde $(m \times m)$-matrix:

$$
V = \begin{pmatrix}
1 & x_1 & x_1^2 & \cdots & x_1^{m-1} \\
1 & x_2 & x_2^2 & \cdots & x_2^{m-1} \\
\vdots & \vdots & \vdots & \ddots & \vdots \\
1 & x_m & x_m^2 & \cdots & x_m^{m-1}
\end{pmatrix}.
$$

By using Exercise 6.4, for the determinant of V it holds:

$$
\det(V) = \prod_{1 \le i < j \le m} (x_j - x_i).
$$

Since all x_i, $i = 1, \ldots, n$ are pairwise different by assumption, the determinant of V does not vanish, and, hence, the matrix V is regular. Thus, the data matrix X has indeed full rank m, and the normal equation is uniquely solvable:

$$
w = \left(X^T \cdot X \right)^{-1} \cdot X^T \cdot y.
$$

Exercise 6.6 (Mean Squared Error)
We first show the bias-variance decomposition of the mean squared error. We enlarge the latter by $\mathbb{E}(w_{lin})$, which yields:

$$
\mathbb{E} \, \| w_{lin} - w \|_2^2 = \mathbb{E} \, \| w_{lin} - \mathbb{E}(w_{lin}) + \mathbb{E}(w_{lin}) - w \|_2^2.
$$

We proceed computations and expand the terms:

$$
\mathbb{E} \left(\| w_{lin} - \mathbb{E}(w_{lin}) \|_2^2 + \| \mathbb{E}(w_{lin}) - w \|_2^2 + 2 \cdot \langle w_{lin} - \mathbb{E}(w_{lin}), \mathbb{E}(w_{lin}) - w \rangle \right).
$$

Due to the linearity of expectation, equation above can be written as

$$
\mathbb{E} \, \| w_{lin} - \mathbb{E}(w_{lin}) \|_2^2 + \| \mathbb{E}(w_{lin}) - w \|_2^2 + 2 \cdot \mathbb{E} \left(\langle w_{lin} - \mathbb{E}(w_{lin}), \mathbb{E}(w_{lin}) - w \rangle \right).
$$

Having a closer look, we conclude that the last summand vanishes:

$$
\mathbb{E} \left(\langle w_{lin} - \mathbb{E}(w_{lin}), \mathbb{E}(w_{lin}) - w \rangle \right) = \langle \underbrace{\mathbb{E}(w_{lin}) - \mathbb{E}(w_{lin})}_{=0}, \mathbb{E}(w_{lin}) - w \rangle.
$$

Hence, the mean squared error simplifies to

$$\mathbb{E} \left\| w_{lin} - w \right\|_2^2 = \left\| \mathbb{E} \left(w_{lin} \right) - w \right\|_2^2 + \mathbb{E} \left\| w_{lin} - \mathbb{E} \left(w_{lin} \right) \right\|_2^2 .$$

It remains to analyze the second term. Note that

$$\left\| w_{lin} - \mathbb{E} \left(w_{lin} \right) \right\|_2^2 = \sum_{i=0}^{m-1} \left((w_{lin})_i - (\mathbb{E} \left(w_{lin} \right))_i \right)^2 ,$$

from where we deduce by using linearity of expectation that

$$\mathbb{E} \left\| w_{lin} - \mathbb{E} \left(w_{lin} \right) \right\|_2^2 = \sum_{i=0}^{m-1} \mathbb{E} \left(\left((w_{lin})_i - (\mathbb{E} \left(w_{lin} \right))_i \right)^2 \right) .$$

Thus, the second term is the sum of diagonal elements of the variance matrix of w_{lin} and equals therefore to its trace:

$$\mathbb{E} \left\| w_{lin} - \mathbb{E} \left(w_{lin} \right) \right\|_2^2 = \text{trace} \left(\text{Var} \left(w_{lin} \right) \right) .$$

Altogether, the bias-variance decomposition of the mean squared error holds:

$$\mathbb{E} \left\| w_{lin} - w \right\|_2^2 = \left\| \mathbb{E} \left(w_{lin} \right) - w \right\|_2^2 + \text{trace} \left(\text{Var} \left(w_{lin} \right) \right) .$$

As an immediate consequence, the bias term vanishes for every unbiased estimator. Under the assumption $X^T \cdot X = n \cdot I$, we calculate the mean squared error of w_{OLS} and w_{ridge}. Since the former is unbiased, it remains to compute the trace of $\text{Var} \left(w_{OLS} \right)$. We recall the formula for the variance of the OLS estimator:

$$\text{Var} \left(w_{OLS} \right) = \sigma^2 \cdot \underbrace{\left(X^T \cdot X \right)^{-1}}_{= \frac{1}{n} \cdot I} = \frac{\sigma^2}{n} \cdot I.$$

Hence, for the mean squared error of w_{OLS} we get:

$$\mathbb{E} \left\| w_{OLS} - w \right\|_2^2 = \text{trace} \left(\text{Var} \left(w_{OLS} \right) \right) = \text{trace} \left(\frac{\sigma^2}{n} \cdot I \right) = \frac{m}{n} \cdot \sigma^2.$$

The ridge estimator is biased, since $\mathbb{E} \left(w_{ridge} \right) = \frac{n}{n+\lambda} \cdot w$. Therefore, the bias term simplifies to

$$\left\| \mathbb{E} \left(w_{ridge} \right) - w \right\|_2^2 = \frac{\lambda^2}{(n+\lambda)^2} \cdot \left\| w \right\|_2^2.$$

Further, the variance of the ridge estimator is given by $\mathrm{Var}\left(w_{ridge}\right) = \frac{n^2}{(n+\lambda)^2} \cdot \mathrm{Var}\left(w_{OLS}\right)$. From here we obtain:

$$\mathrm{trace}\left(\mathrm{Var}\left(w_{ridge}\right)\right) = \frac{n^2}{(n+\lambda)^2} \cdot \mathrm{trace}\left(\mathrm{Var}\left(w_{OLS}\right)\right) = \frac{n^2}{(n+\lambda)^2} \cdot \frac{m}{n} \cdot \sigma^2 = \frac{n \cdot m}{(n+\lambda)^2} \cdot \sigma^2.$$

Finally we have derived the formula of the mean squared error of w_{ridge}:

$$\mathbb{E}\left\|w_{ridge} - w\right\|_2^2 = \frac{\lambda^2}{(n+\lambda)^2} \cdot \|w\|_2^2 + \frac{n \cdot m}{(n+\lambda)^2} \cdot \sigma^2.$$

Wee see that for small values of λ the variance term dominates the bias. For large λ's, however, the mean square error is controlled by the size of the true weights. This gives rise for the choice of λ, which minimizes the mean square error. Solving the optimization problem:

$$\min_{\lambda} \frac{\lambda^2 \cdot \|w\|_2^2 + n \cdot m \cdot \sigma^2}{(n+\lambda)^2}$$

gives us the Tikhonov parameter:

$$\lambda = \frac{m \cdot \sigma^2}{\|w\|_2^2}.$$

10.7 Sparsity

Exercise 7.1 (Zero Norm)
We show that $\|\cdot\|_0$ satisfies all norm properties except of the absolute homogeneity. Let us start with the positive definiteness. Clearly, $\|w\|_0 \geq 0$ for all $w \in \mathbb{R}^n$. Further, the zero vector has by definition no nonzero elements and, hence, $\|0\|_0 = 0$. On the other hand, any nonzero vector has at least one element which is not equal to zero, i.e. for any $w \neq 0$ we have $\|w\|_0 > 0$. In order to show the triangle inequality, take any arbitrary two vectors $v, w \in \mathbb{R}^n$. If $v_j + w_j \neq 0$ for an index j, then $v_j \neq 0$ or $w_j \neq 0$. Hence, we have:

$$\|v + w\|_0 = \#\left\{j \mid v_j + w_j \neq 0\right\} \leq \#\left\{j \mid v_j \neq 0\right\} + \#\left\{j \mid w_j \neq 0\right\} = \|v\|_0 + \|w\|_0.$$

We show that $\|\cdot\|_0$ is not absolutely homogeneous, namely it holds for $\alpha \in \mathbb{R}$ and $w \in \mathbb{R}^n$:

$$\|\alpha \cdot w\|_0 = |\alpha| \cdot \|w\|_0 \text{ if and only if } \alpha \in \{0, \pm 1\} \text{ or } w = 0.$$

If $\alpha = 0$ or $w = 0$, the equality trivially holds due to the positive definiteness. If $\alpha \neq 0$, a vector w has the same non-zero entries as $\alpha \cdot w$. Hence, their zero norms coincide:

$$\|w\|_0 = \|\alpha \cdot w\|_0.$$

By assuming the absolute homogeneity to hold, we obtain:

$$\|w\|_0 = |\alpha| \cdot \|w\|_0.$$

If $w \neq 0$, we may divide this equality by $\|w\|_0$ to get $|\alpha| = 1$.

Exercise 7.2 (Spark)
The first step for showing optimality is to check whether the vector $w = (0, 12, 0, 6, 0, 0)^T$ is feasible, i.e. $X \cdot w = y$ holds. This is indeed the case:

$$\begin{pmatrix} 1 & -1 & 1 & 0 & 0 & 0 \\ 1 & 0 & -1 & 1/2 & 1/2 & 0 \\ 1 & 1 & 0 & -1 & 0 & 0 \\ 1 & 1/3 & 0 & 1/3 & -1 & 1/3 \\ 1 & 1/3 & 1/3 & 1/3 & 0 & -1 \end{pmatrix} \cdot \begin{pmatrix} 0 \\ 12 \\ 0 \\ 6 \\ 0 \\ 0 \end{pmatrix} = \begin{pmatrix} -12 \\ 3 \\ 6 \\ 6 \\ 6 \end{pmatrix}.$$

Further, let us examine the sufficient condition for optimality:

$$\|w\|_0 < \frac{\text{spark}(X)}{2},$$

which requires to compute the spark of the matrix X. For that, we have to go through the combinations of the column vectors. Clearly, any two columns of X are linearly independent. Let us proceed by considering the subsets of three columns. Note that there are 20 combinations of three columns we can choose out of X. We inspect e.g. the first three columns by using the Gaussian elimination:

$$\begin{matrix} 1 & -1 & 1 \\ 1 & 0 & -1 \\ 1 & 1 & 0 \\ 1 & 1/3 & 0 \\ 1 & 1/3 & 1/3 \end{matrix} \sim \begin{matrix} 1 & 0 & -1 \\ 0 & 1 & -2 \\ 0 & 0 & 3 \\ 0 & 0 & 5/3 \\ 0 & 0 & 2 \end{matrix} \sim \begin{matrix} 1 & 0 & 0 \\ 0 & 1 & 0 \\ 0 & 0 & 1 \\ 0 & 0 & 0 \\ 0 & 0 & 0 \end{matrix}.$$

Hence, they are linearly independent. The other 19 combinations yield the same result. We conclude that spark $(X) \geq 4$. Similar calculations provide that any of four columns are linearly independent, thus, spark $(X) \geq 5$. However, the five last columns are linearly dependent:

$$
\begin{pmatrix}
-1 & 1 & 0 & 0 & 0 \\
0 & -1 & 1/2 & 1/2 & 0 \\
1 & 0 & -1 & 0 & 0 \\
1/3 & 0 & 1/3 & -1 & 1/3 \\
1/3 & 1/3 & 1/3 & 0 & -1
\end{pmatrix}
\sim
\begin{pmatrix}
1 & 0 & 0 & 0 & 1 \\
0 & 1 & 0 & 0 & 1 \\
0 & 0 & 1 & 0 & -1 \\
0 & 0 & 0 & 1 & -1 \\
0 & 0 & 0 & 0 & 0
\end{pmatrix}.
$$

This provides that spark $(X) = 5$. Hence, w solves the optimization problem $(\mathcal{P}_0)$, as

$$
\|w\|_0 = 2 < \frac{5}{2} = \frac{\text{spark}(X)}{2}.
$$

Exercise 7.3 (Null Space Property)
We analyze the null space property of the matrix

$$
X = \begin{pmatrix}
1 & 0 & 1 & 0 \\
0 & 1 & 1 & 0 \\
0 & 1 & 0 & 1
\end{pmatrix}.
$$

Solving the homogeneous system $X \cdot u = 0$ provides that the null space of X can be spanned by the vector $u = (1, 1, -1, -1)^T$. It holds for an arbitrary index subset $S \subset \{1, 2, 3, 4\}$:

$$
\|u_S\|_1 = |S|.
$$

Hence, whenever the cardinality s of the set S is greater than or equal to the cardinality $4 - s$ of S^c, the null space property for the vector $u \in \text{null}(X)$ does not hold. The matrix X fulfills the null space property if and only if $s \in \{0, 1\}$.

Exercise 7.4 (Laplace Distribution)
We want to evaluate:

$$
\mathbb{E}(Z) = \int_{-\infty}^{\infty} z \cdot p(z) \, dz = \int_{-\infty}^{\infty} z \cdot \frac{1}{2\tau} \cdot e^{-\frac{|z-\mu|}{\tau}} \, dz.
$$

We substitute $x = z - \mu$, which provides:

$$
\int_{-\infty}^{\infty} (x + \mu) \cdot \frac{1}{2\tau} \cdot e^{-\frac{|x|}{\tau}} \, dx = \int_{-\infty}^{\infty} x \cdot \frac{1}{2\tau} \cdot e^{-\frac{|x|}{\tau}} \, dx + \int_{-\infty}^{\infty} \mu \cdot \frac{1}{2\tau} \cdot e^{-\frac{|x|}{\tau}} \, dx.
$$

Let us inspect the first integral and simplify it:

$$\int_{-\infty}^{\infty} x \cdot \frac{1}{2\tau} \cdot e^{-\frac{|x|}{\tau}} \, dx = \frac{1}{2\tau} \cdot \left[\int_{-\infty}^{0} x \cdot e^{\frac{x}{\tau}} \, dx + \int_{0}^{\infty} x \cdot e^{-\frac{x}{\tau}} \, dx \right].$$

Integrating by parts yields:

$$\frac{1}{2\tau} \cdot \left[\int_{-\infty}^{0} x \cdot e^{\frac{x}{\tau}} \, dx + \int_{0}^{\infty} x \cdot e^{-\frac{x}{\tau}} \, dx \right] = \frac{1}{2\tau} \cdot \left(-\tau^2 + \tau^2 \right) = 0.$$

It remains to evaluate the second integral, which can be written as

$$\frac{1}{2\tau} \cdot \int_{-\infty}^{\infty} \mu \cdot e^{-\frac{|x|}{\tau}} \, dx = \frac{\mu}{2\tau} \cdot \left[\int_{-\infty}^{0} e^{\frac{x}{\tau}} \, dx + \int_{0}^{\infty} e^{-\frac{x}{\tau}} \, dx \right].$$

Straightforward computations yield:

$$\frac{\mu}{2\tau} \cdot \left[\int_{-\infty}^{0} e^{\frac{x}{\tau}} \, dx + \int_{0}^{\infty} e^{-\frac{x}{\tau}} \, dx \right] = \frac{\mu}{2\tau} \cdot (\tau + \tau) = \mu.$$

Hence, we conclude for the expectation:

$$\mathbb{E}(Z) = 0 + \mu = \mu.$$

We compute the variance of Z by defining the random variable $X = Z - \mu$. Then, we have:

$$\mathbb{E}(X) = \mathbb{E}(Z) - \mu = 0.$$

Together with linearity of the expectation this yields:

$$\text{Var}(Z) = \mathbb{E}\left(Z^2\right) - \mathbb{E}(Z)^2 = \mathbb{E}\left((X + \mu)^2\right) - (\mathbb{E}(X) + \mu)^2$$

$$= \mathbb{E}\left(X^2\right) + 2 \cdot \underbrace{\mathbb{E}(X)}_{=0} \cdot \mu + \mu^2 - \underbrace{\mathbb{E}(X)^2}_{=0} - 2 \cdot \underbrace{\mathbb{E}(X)}_{=0} \cdot \mu - \mu^2 = \mathbb{E}\left(X^2\right).$$

Therefore, it remains to evaluate the integral

$$\frac{1}{2\tau} \int_{-\infty}^{\infty} x^2 \cdot e^{-\frac{|x|}{\tau}} \, dx = \frac{1}{2\tau} \cdot \left(\int_{-\infty}^{0} x^2 \cdot e^{\frac{x}{\tau}} \, dx + \int_{0}^{\infty} x^2 \cdot e^{-\frac{x}{\tau}} \, dx \right).$$

Let us compute the first integral. Integration by parts provides:

$$\int_{-\infty}^{0} x^2 \cdot e^{\frac{x}{\tau}}\, dx = 0 - 2 \cdot \tau \cdot \underbrace{\int_{-\infty}^{0} x \cdot e^{\frac{x}{\tau}}\, dx}_{=-\tau^2} = 2 \cdot \tau \cdot \tau^2 = 2\tau^3.$$

Similar computations provide

$$\int_{0}^{\infty} x^2 \cdot e^{-\frac{x}{\tau}}\, dx = 0 + 2 \cdot \tau \cdot \underbrace{\int_{0}^{\infty} x \cdot e^{-\frac{x}{\tau}}\, dx}_{=\tau^2} = 2 \cdot \tau^3.$$

Altogether, we get:

$$\mathrm{Var}(Z) = \mathbb{E}\left(X^2\right) = \frac{1}{2\tau} \cdot \left(2 \cdot \tau^3 + 2 \cdot \tau^3\right) = 2\tau^2.$$

Exercise 7.5 (Lasso)
We have to show the equivalence of the convex optimization problems:

$$\min_{w} \; \frac{1}{2} \cdot \|y - X \cdot w\|_2^2 + \lambda \cdot \|w\|_1 \qquad\qquad (\mathcal{L}asso)$$

and

$$\min_{w} \; \frac{1}{2} \cdot \|y - X \cdot w\|_2^2 \quad \text{s.t.} \quad \|w\|_1 \le s. \qquad\qquad (\mathcal{V}\text{-}\mathcal{L}asso)$$

For brevity we denote:

$$f(w) = \frac{1}{2} \cdot \|y - X \cdot w\|_2^2, \quad g(w) = \|w\|_1.$$

Let us first show that any solution w of $(\mathcal{V} - \mathcal{L}asso)$ also solves $(\mathcal{L}asso)$. By introducing the Lagrange multiplier $\mu \in \mathbb{R}$ for the inequality constraint, w satisfies the optimality condition for $(\mathcal{V} - \mathcal{L}asso)$, see e.g. Rockafellar (1970):

$$\text{Stationarity: } 0 \in \nabla f(w) + \mu \cdot \partial g(w),$$

$$\text{Primal feasibility: } \|w\|_1 \le s,$$

$$\text{Dual feasibility: } \mu \ge 0,$$

$$\text{Complementary slackness: } \mu \cdot (\|w\|_1 - s) = 0.$$

Here, $\partial g\,(w)$ denotes the convex subdifferential of the Manhattan norm. If we set $\lambda = \mu$, then stationarity provides the optimality condition for ($\mathcal{L}asso$), and, hence, w solves the latter due to its convexity. We prove the other direction in a similar way. Let w be a solution of ($\mathcal{L}asso$). Then, w must fulfill:

$$0 \in \nabla f\,(w) + \lambda \cdot \partial g\,(w).$$

We can choose $s = \|w\|_1$ and $\mu = \lambda$. Apparently, w and μ are primal and dual feasible, respectively. Furthermore, $\|w\|_1 - s = 0$, i.e. the constraint is active, and the complementary slackness holds. Additionally, w is stationary due to the optimality condition for ($\mathcal{L}asso$). Altogether, we conclude that the optimality condition holds and, hence, w solves ($\mathcal{V} - \mathcal{L}asso$).

Exercise 7.6 (ISTA)
We start by computing the matrix

$$X^T \cdot X = \begin{pmatrix} 3 & -2 \\ 2 & 0 \\ 6 & 8 \end{pmatrix} \cdot \begin{pmatrix} 3 & 2 & 6 \\ -2 & 0 & 8 \end{pmatrix} = \begin{pmatrix} 13 & 6 & 2 \\ 6 & 4 & 12 \\ 2 & 12 & 100 \end{pmatrix}.$$

Due to Exercise 3.2, the singular values σ of X are the square roots of the eigenvalues λ of $X^T \cdot X$. The latter are characterized by

$$\det\left(X^T \cdot X - \lambda \cdot I\right) = 0.$$

Straightforward calculations provide the square roots of the eigenvalues:

$$\sqrt{\lambda_1} = 10.0777, \quad \sqrt{\lambda_2} = 3.9292, \quad \sqrt{\lambda_3} = 0.$$

We conclude that

$$\sigma_{\max}\,(X) = 10.0777,$$

and, hence, the Lipschitz constant for the gradients of f is

$$L = \sigma_{\max}^2\,(X) = 101.5600.$$

The ($\mathcal{IST\!A}$) update reads as

$$w\,(t+1) = T_{\frac{4}{101.56}}\left(w\,(t) - \frac{1}{101.56} \cdot X^T \cdot (X \cdot w(t) - y)\right).$$

To clarify the algorithm, let us demonstrate how to determine the new weights after the first iteration. We start by computing

$$w\,(1) - \frac{1}{101.56} \cdot X^T \cdot (X \cdot w(1) - y) = \begin{pmatrix} 0.744 \\ 0.8031 \\ 0.2517 \end{pmatrix}.$$

It remains to apply the soft-thresholding operator to the resulting vector. Since $\lambda = 4$ and $L = 101.56$, we get $\frac{\lambda}{L} \approx 0.0394$. Note that all elements of the vector $(0.744, 0.8031, 0.2517)^T$ are positive and strictly greater than 0.0394. Therefore, the soft-thresholding operator simplifies to

$$T_{\frac{4}{101.56}} \left(w\,(t) - \frac{1}{101.56} \cdot X^T \cdot (X \cdot w(t) - y) \right) = \begin{pmatrix} 0.744 \\ 0.8031 \\ 0.2517 \end{pmatrix} - 0.0394 \cdot e = \begin{pmatrix} 0.7046 \\ 0.7637 \\ 0.2123 \end{pmatrix}.$$

The new weights are given by

$$w\,(2) = \begin{pmatrix} 0.7046 \\ 0.7637 \\ 0.2123 \end{pmatrix}.$$

By using the Python code, we see that after 13 iterations the weights become sparse as the second entry vanishes:

$$w\,(13) = \begin{pmatrix} -0.123 \\ 0 \\ 0.3401 \end{pmatrix}.$$

The sparsity pattern prevails during further iterations. Moreover, after 14 iterations the difference of the objective function values only slightly changes:

$$f\,(w\,(15)) - f\,(w\,(14)) < 0.0001.$$

10.8   Neural Networks

Exercise 8.1 (Neural Network)
In order to distribute the resources for advertisements efficiently, we calculate the probability of a purchase response for each new customer. The model has one layer, activated by the sigmoid function and based on the features age, income, and number

of previous purchases. Further, the weights of these inputs have already been estimated:

$$w_0 = 0, \quad w_1 = -0.1, \quad w_2 = 0.6, \quad w_3 = 0.7.$$

We compute the output of the input layer $z = w^T \cdot X$, which will be then passed into the sigmoid activation function $f_S(z) = \frac{1}{1+e^{-z}}$. For customer 1 we get:

$$z_1 = w^T \cdot x_1 = -0.1 \cdot 20 + 0.6 \cdot 6 + 0.7 \cdot 1 = 2.3.$$

This yields to the following purchase probability:

$$f_S(2.3) = \frac{1}{1 + e^{-2.3}} \approx 0.9089.$$

Hence, it is very likely that customer 1 reacts with a purchase to an advertisement. In contrast, customer 2 will react with a purchase to an advertisement as likely as he will not, since

$$z_2 = -0.1 \cdot 30 + 0.6 \cdot 5 + 0.7 \cdot 0 = 0,$$

and

$$f_S(0) = \frac{1}{1 + e^0} = 0.5.$$

Similar calculations for customer 3 provide:

$$z_3 = w^T \cdot x_3 = -0.1 \cdot 40 + 0.6 \cdot 1 + 0.7 \cdot 3 = -1.3,$$

and

$$f_S(-1.3) = \frac{1}{1 + e^{1.3}} \approx 0.2142.$$

Let us briefly interpret these results. Customer 1 has a very high income compared to customer 3. Further, the latter is older, and age has negative influence on the purchase response due to the minus sign of w_1. The comparison also shows that a decision solely based on the purchase history, may lead to wrong conclusion. Even though customer 3 bought previously more frequently than the others, we derived the lowest response probability. This shows that exploiting the information given by the combination of features, is indeed rather efficient. From the economic point of view, even a low response probability could justify an advertisement if e.g. the cost of the latter are very low, while the possible margin of the sold product is very high. Yet, the predicted purchase response provides a precise decision foundation.

Exercise 8.2 (Sigmoid Activation Function)
We solve the initial value problem. Separation of variables provides:

$$\frac{\mathrm{d}f}{f \cdot (1 - f)} = \mathrm{d}z.$$

Integrating both sides of the equation yields:

$$\int \frac{\mathrm{d}f}{f \cdot (1 - f)} = \int \mathrm{d}z,$$

and, hence:

$$\ln f - \ln (1 - f) = z + C,$$

where C is a constant. Let us simplify this equation:

$$f(z) = \frac{e^C}{e^C + e^{-z}}.$$

It remains to apply the initial value condition $f(0) = \frac{1}{2}$:

$$\frac{1}{2} = \frac{e^C}{e^C + 1},$$

and, hence:

$$e^C = 1.$$

Substituting the constant into the equation above, we conclude that the sigmoid activation function solves the initial value problem, i.e.

$$f(z) = \frac{1}{1 + e^{-z}}.$$

Exercise 8.3 (Logistic Distribution)
We evaluate the integral:

$$\mathbb{P}(\varepsilon \leq z) = \int_{-\infty}^{z} p(z) \, \mathrm{d}z = \int_{-\infty}^{z} \frac{e^{-z}}{\left(1 + e^{-z}\right)^2} \, \mathrm{d}z.$$

By substituting $u = 1 + e^{-z}$, we have:

$$\int_{-\infty}^{z} \frac{e^{-z}}{\left(1 + e^{-z}\right)^2} \, dz = -\int_{\infty}^{u} \frac{1}{u^2} \, du = \int_{u}^{\infty} \frac{du}{u^2} = -\frac{1}{u}\bigg|_{u}^{\infty} = \frac{1}{u} = \frac{1}{1 + e^{-z}} = f_S(z).$$

Hence, the sigmoid activation function is indeed a cumulative distribution function of a random variable following the standard logistic distribution. Let us calculate its expectation:

$$\mathbb{E}(\epsilon) = \int_{-\infty}^{\infty} z \cdot p(z) \, dz = \int_{-\infty}^{\infty} z \cdot \frac{e^{-z}}{\left(1 + e^{-z}\right)^2} \, dz.$$

Again, we apply a substitution of variables $x = \frac{1}{1+e^{-z}}$, which provides:

$$\int_{-\infty}^{\infty} z \cdot \frac{e^{-z}}{\left(1 + e^{-z}\right)^2} \, dz = \int_{0}^{1} \ln\left(\frac{x}{1 - x}\right) \, dx = \int_{0}^{1} \ln x \, dx - \int_{0}^{1} \ln(1 - x) \, dx.$$

We deduce by substituting $y = 1 - x$:

$$\int_{0}^{1} \ln(1 - x) \, dx = -\int_{1}^{0} \ln y \, dy = \int_{0}^{1} \ln y \, dy.$$

Therefore, we conclude that the difference of the integrals vanishes, i.e.

$$\int_{0}^{1} \ln x \, dx - \int_{0}^{1} \ln(1 - x) \, dx = \int_{0}^{1} \ln x \, dx - \int_{0}^{1} \ln y \, dy = 0.$$

Thus, a random variable following the standard logistic distribution has zero expectation. The variance needs some thorough computation. At first, we state due to symmetry:

$$\text{Var}(\epsilon) = \int_{-\infty}^{\infty} z^2 \cdot \frac{e^{-z}}{\left(1 + e^{-z}\right)^2} \, dz = 2 \cdot \int_{0}^{\infty} z^2 \cdot \frac{e^{-z}}{\left(1 + e^{-z}\right)^2} \, dz.$$

Next, we analyze $\frac{e^{-z}}{(1+e^{-2z})^2}$ by applying polynomial division, which yields:

$$e^{-z} : \left(1 + 2 \cdot e^{-z} + e^{-2z}\right) = e^{-z} - 2 \cdot e^{-2z} + 3 \cdot e^{-3z} - 4 \cdot e^{-4z} + 5 \cdot e^{-5z} + \dots.$$

Hence, we can replace the fraction by a series:.

$$\frac{e^{-z}}{\left(1 + e^{-2z}\right)^2} = \sum_{n=1}^{\infty} (-1)^{n+1} \cdot n \cdot e^{-nz}.$$

Finally, the original integral becomes:

$$2 \cdot \int_0^\infty z^2 \cdot \sum_{n=1}^\infty (-1)^{n+1} \cdot n \cdot e^{-nz} \, dz = 2 \cdot \sum_{n=1}^\infty (-1)^{n+1} \cdot n \cdot \int_0^\infty z^2 \cdot e^{-nz} \, dz,$$

where integral and summation were interchanged. Let us integrate by parts with $u = z^2$ and $v' = e^{-nz}$:

$$\int_0^\infty z^2 \cdot e^{-nz} \, dz = -\frac{1}{n} \cdot z^2 \cdot e^{-nz} \Big|_0^\infty + \frac{2}{n} \cdot \int_0^\infty z \cdot e^{-nz} \, dz.$$

Clearly, the first term vanishes, while we apply integration by parts for the second term. We set $u = z$ and $v' = e^{-nz}$:

$$\int_0^\infty z \cdot e^{-nz} \, dz = -\frac{1}{n} \cdot z \cdot e^{-nz} \Big|_0^\infty + \frac{1}{n} \cdot \int_0^\infty e^{-nz} \, dz = \frac{1}{n^2}.$$

We conclude that

$$\int_0^\infty z^2 \cdot e^{-nz} \, dz = \frac{2}{n^3}.$$

We derive for the variance:

$$2 \cdot \sum_{n=1}^\infty (-1)^{n+1} \cdot n \cdot \int_0^\infty z^2 \cdot e^{-nz} \, dz = 4 \cdot \sum_{n=1}^\infty (-1)^{n+1} \cdot \frac{1}{n^2}.$$

Next, we take a look at the sum

$$4 \cdot \sum_{n=1}^\infty (-1)^{n+1} \cdot \frac{1}{n^2} = 4 \cdot \left(\frac{1}{1^2} - \frac{1}{2^2} + \frac{1}{3^2} - \frac{1}{4^2} + \frac{1}{5^2} + \ldots \right).$$

Thorough thinking leads to the observation that this sum is similar to the famous series:

$$\sum_{n=1}^\infty \frac{1}{n^2} = \frac{\pi^2}{6}.$$

Let us split this series in the odd and even components:

$$\left(\frac{1}{1^2} + \frac{1}{3^2} + \frac{1}{5^2} + \ldots \right) + \left(\frac{1}{2^2} + \frac{1}{4^2} + \frac{1}{6^2} + \ldots \right).$$

The latter can be written as

$$\frac{1}{2^2}+\frac{1}{4^2}+\frac{1}{6^2}+\ldots=\frac{1}{1^2\cdot 2^2}+\frac{1}{2^2\cdot 2^2}+\frac{1}{3^2\cdot 2^2}+\ldots=\frac{1}{2^2}\left(\frac{1}{1^2}+\frac{1}{2^2}+\frac{1}{3^2}+\ldots\right)=\frac{\pi^2}{24}.$$

Hence, we are able to compute the sum of the odd components:

$$\frac{1}{1^2}+\frac{1}{3^2}+\frac{1}{5^2}+\ldots=\frac{\pi^2}{6}-\frac{\pi^2}{24}=\frac{3\cdot\pi^2}{24}.$$

Note that our series is the difference of the odd and even terms:

$$4\cdot\sum_{n=1}^{\infty}(-1)^{n+1}\cdot\frac{1}{n^2}=4\cdot\left(\frac{3\cdot\pi^2}{24}-\frac{\pi^2}{24}\right)=\frac{\pi^2}{3}.$$

Thus, we derived that the variance of the standard logistic distribution is $\pi^2/3$.

Exercise 8.4 (Latent-Variable Model)

We show that the logistic regression can be derived via a latent formulation. Due to $y^*=x^T\cdot w+\varepsilon$, the classification of a newcomer is stochastic. Therefore, let us write y explicitly in terms of the random error ε, i.e.

$$y=\begin{cases}1,\text{ if }\varepsilon\geq -x^T\cdot w,\\0,\text{ else.}\end{cases}$$

Therefore, we obtain:

$$\mathbb{P}\left(y=1\,|\,x\right)=\mathbb{P}\left(\varepsilon\geq -x^T\cdot w\right).$$

We recall from Exercise 8.3 that the sigmoid activation function is the cumulative distribution function of the logistic random variable ε, and continue:

$$1-\mathbb{P}\left(\varepsilon\leq -x^T\cdot w\right)=1-f_S\left(-x^T\cdot w\right)=1-\frac{1}{1+e^{x^T\cdot w}}=\frac{e^{x^T\cdot w}}{1+e^{x^T\cdot w}}=\frac{1}{1+e^{-x^T\cdot w}}.$$

The probability to label a newcomer x by 1 can be thus given in terms of the sigmoid activation function:

$$\mathbb{P}\left(y=1\,|\,x\right)=f_S\left(x^T\cdot w\right).$$

Exercise 8.5 (Cross-Entropy)

We compute the gradient of $H_i(w)$. From Exercise 8.3 we already know:

$$f_S'(z) = f_S(z) \cdot (1 - f_S(z)).$$

Therefore, applying the chain rule yields for all $i = 1, \ldots, n$:

$$\nabla H_i(w) = -y_i \cdot \frac{1}{f_S\left(x_i^T \cdot w\right)} \cdot f_S'\left(x_i^T \cdot w\right) \cdot x_i$$

$$- (1 - y_i) \cdot \frac{1}{1 - f_S\left(x_i^T \cdot w\right)} \cdot f_S'\left(x_i^T \cdot w\right) \cdot x_i$$

$$= -y_i \cdot \frac{1}{f_S\left(x_i^T \cdot w\right)} \cdot f_S\left(x_i^T \cdot w\right) \cdot \left(1 - f_S\left(x_i^T \cdot w\right)\right) \cdot x_i$$

$$- (1 - y_i) \cdot \frac{1}{1 - f_S\left(x_i^T \cdot w\right)} \cdot f_S\left(x_i^T \cdot w\right) \cdot \left(1 - f_S\left(x_i^T \cdot w\right)\right) \cdot x_i$$

$$= \left(f_S\left(x_i^T \cdot w\right) - y_i\right) \cdot x_i.$$

In order to derive the Hesse matrix of $H_i(w)$, we calculate for all $i = 1, \ldots, n$:

$$\nabla^2 H_i(w) = \nabla\left(\nabla^T H_i(w)\right) = \nabla\left(f_S\left(x_i^T \cdot w\right) - y_i\right) \cdot x_i^T$$

$$= f_S'\left(x_i^T \cdot w\right) \cdot x_i \cdot x_i^T = f_S(z) \cdot (1 - f_S(z)) \cdot x_i \cdot x_i^T.$$

The latter also shows the convexity of the cross-entropy.

Exercise 8.6 (Multilayer Perceptron)

We solve the XOR problem by a neural network with one hidden layer as represented in Fig. 8.7. Since we have two hidden neurons, we have two vectors of weights

$$w^1 = \left(-\frac{1}{3}, \frac{1}{2}, -\frac{1}{2}\right)^T, \quad w^2 = \left(-\frac{1}{3}, -\frac{1}{2}, \frac{1}{2}\right)^T.$$

Let us calculate the outputs z_1, z_2 of each hidden neuron:

$$z_1 = f_T\left(\sum_{j=1}^{2} w_j^1 \cdot x_j - \frac{1}{3}\right), \quad z_2 = f_T\left(\sum_{j=1}^{2} w_j^2 \cdot x_j - \frac{1}{3}\right).$$

Hence, for the different pairs of inputs we get the following outputs of z_1:

$$(1, 1) : f_T\left(-\tfrac{1}{3}\right) = 0, \ (1, 0) : f_T\left(\tfrac{1}{6}\right) \ = 1,$$

$$(0, 1) : f_T\left(-\tfrac{5}{6}\right) = 0, \ (0, 0) : f_T\left(-\tfrac{1}{3}\right) = 0.$$

For the second hidden neuron we get the outputs z_2 in dependence of the input:

$$(1, 1) : f_T\left(-\tfrac{1}{3}\right) = 0, \ (1, 0) : f_T\left(-\tfrac{5}{6}\right) = 0,$$

$$(0, 1) : f_T\left(\tfrac{1}{6}\right) \ = 1, \ (0, 0) : f_T\left(-\tfrac{1}{3}\right) = 0.$$

The output layer y is given by the sum of the hidden layers' outputs, i.e. $y = z_1 + z_2$. Finally, we can compute the outputs in dependence of the inputs:

$$(1, 1) : 0 + 0 = 0, \ (1, 0) : 1 + 0 = 1,$$

$$(0, 1) : 0 + 1 = 1, \ (0, 0) : 0 + 0 = 0.$$

Indeed, the multilayer perceptron from Fig. 8.7 outputs $y = 1$ only for the pairs of inputs $x = (1, 0)$ and $x = (0, 1)$, therefore, it solves the XOR problem.

10.9   Decision Trees

Exercise 9.1 (Binary Classification)
Consider the following binary class assignment for the training set $\{0, 1\}^3$:

Object	$(0,0,0)$	$(0,0,1)$	$(0,1,0)$	$(0,1,1)$	$(1,0,0)$	$(1,0,1)$	$(1,1,0)$	$(1,1,1)$
Class	C_{yes}	C_{no}	C_{no}	C_{yes}	C_{yes}	C_{no}	C_{no}	C_{yes}

Let a decision tree D select the test T_1 at the root, and then successively the tests T_2, T_3 at the subtrees' roots. Every path of this decision tree leads to a unique binary vector $x \in \{0, 1\}^3$. At the last step we set $D(x) = C_{yes}$ or C_{no} to guarantee the correct classification, see Fig. 10.1. On the training set $\{0, 1\}^3$ endowed with the above binary class assignment the misclassification rate of the decision tree D vanishes. For its average external path length holds $\rho(D) = 3$. Obviously, the decision tree D minimizes the average external path length for the proposed binary class assignment due to symmetry. In order to correctly classify an object by means of a decision tree, all of its three entries need to be thus checked

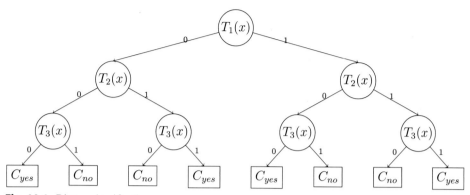

Fig. 10.1 Binary classification

on average. It is not hard to see that the proposed class assignment maximizes the minimal average external path length among all decision trees with zero misclassification rate.

Exercise 9.2 (Minimal External Path Length)
The identification decision trees with the minimal external path lengths $f(n)$ on n-element sets are given in Figs. 10.2, 10.3, 10.4, 10.5, 10.6, 10.7, 10.8, and 10.9 for $n = 1, \ldots, 8$. Note that here all 1- and 3-element subsets are available as tests. At the leafs we give the lengths of the corresponding external paths, which sum up to $f(n)$.

Exercise 9.3 (Matching Problem)
For a given matching problem instance 3DM($\mathcal{A}$, $\mathcal{M}$) we construct a corresponding exact cover instance EC3($\mathcal{Y}$, $\mathcal{E}$), so that solving the latter will provide us with a matching for the former. For that, we set:

$$\mathcal{Y} = \mathcal{A}, \quad \mathcal{E} = \{\{a_1, a_2, a_3\} \mid (a_1, a_2, a_3) \in \mathcal{M}\}.$$

By construction, we have for all $E \in \mathcal{E}$:

$$E \subset \mathcal{Y} \quad \text{and} \quad |E| = 3.$$

Solving the exact cover problem EC3($\mathcal{Y}$, $\mathcal{E}$) gives us a subcollection $\mathcal{E}^* \subset \mathcal{E}$, such that each element $y \in \mathcal{Y}$ is contained in exactly one of its subsets $E \in \mathcal{E}^*$. In particular, it holds $|\mathcal{E}^*| = |\mathcal{Y}|$. The exact cover defines a matching by setting

$$\mathcal{M}^* = \{(a_1, a_2, a_3) \mid \{a_1, a_2, a_3\} \in \mathcal{E}^*\}.$$

Moreover, we have $|\mathcal{M}^*| = |\mathcal{A}|$. Overall, we have shown that the matching problem 3DM($\mathcal{A}$, $\mathcal{M}$) is reducible to the exact cover problem EC3($\mathcal{Y}$, $\mathcal{E}$) in polynomial time. Since

Fig. 10.2 $f(1) = 0$

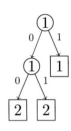

Fig. 10.3 $f(2) = 2$

Fig. 10.4 $f(3) = 5$

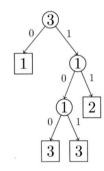

Fig. 10.5 $f(4) = 9$

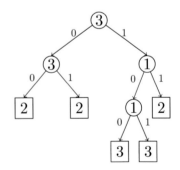

Fig. 10.6 $f(5) = 12$

Fig. 10.7 $f(6) = 16$

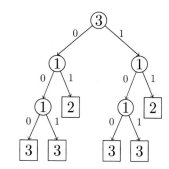

Fig. 10.8 $f(7) = 21$

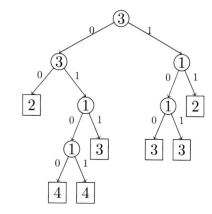

Fig. 10.9 $f(8) = 25$

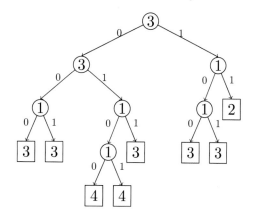

the former is known to be NP-complete, we conclude that the exact cover problem is also NP-complete.

Exercise 9.4 (Suboptimality of Iterative Dichotomizer)
For the set of classified objects we write:

$$\mathcal{X} = \{x_1, x_2, x_3, x_4\},$$

where

$$x_1 = (1, 1, 1)^T, \quad x_2 = (1, 1, 0)^T, \quad x_3 = (1, 0, 0)^T, \quad x_4 = (0, 0, 1)^T.$$

Moreover, we have:

$$C_{yes} = \{x_1, x_3\}, \quad C_{no} = \{x_2, x_4\}.$$

For any $j = 1, 2, 3$ it is possible to test the j-th entry of $x \in \mathcal{O} = \{0, 1\}^3$:

$$T_j(x) = \begin{cases} 0, & \text{if } x_j = 0, \\ 1, & \text{if } x_j = 1. \end{cases}$$

Let us recursively apply the iterative dichotomizer. First, we split the subset $S = \mathcal{X}$:

- Test T_1 gives the split

$$L_1 = \{x_4\}, \quad R_1 = \{x_1, x_2, x_3\}.$$

The corresponding gain is

$$G_1(S) = \varepsilon_1(S) - \left(\frac{|L_1|}{|S|} \cdot \varepsilon_1(L_1) + \frac{|R_1|}{|S|} \cdot \varepsilon_1(R_1) \right) = \frac{2}{4} - \left(\frac{1}{4} \cdot 0 + \frac{3}{4} \cdot \frac{1}{3} \right) = \frac{1}{4}.$$

- Test T_2 gives the split

$$L_2 = \{x_3, x_4\}, \quad R_1 = \{x_1, x_2\}.$$

The corresponding gain is

$$G_2(S) = \varepsilon_1(S) - \left(\frac{|L_2|}{|S|} \cdot \varepsilon_1(L_2) + \frac{|R_2|}{|S|} \cdot \varepsilon_1(R_2) \right) = \frac{2}{4} - \left(\frac{2}{4} \cdot \frac{1}{2} + \frac{2}{4} \cdot \frac{1}{2} \right) = 0.$$

- Test T_3 gives the split

$$L_3 = \{x_2, x_3\}, \quad R_1 = \{x_1, x_4\}.$$

The corresponding gain is

$$G_3(S) = \varepsilon_1(S) - \left(\frac{|L_3|}{|S|} \cdot \varepsilon_1(L_3) + \frac{|R_3|}{|S|} \cdot \varepsilon_1(R_3)\right) = \frac{2}{4} - \left(\frac{2}{4} \cdot \frac{1}{2} + \frac{2}{4} \cdot \frac{1}{2}\right) = 0.$$

We choose the splitting by the test T_1. The data subset L_1 ends up in a leaf C_{no}. Analogously, we have to split $S = R_1$ by the remaining tests:

- Test T_2 gives the split

$$L_2 = \{x_3\}, \quad R_2 = \{x_1, x_2\}.$$

The corresponding gain is

$$G_2(S) = \varepsilon_1(S) - \left(\frac{|L_2|}{|S|} \cdot \varepsilon_1(L_2) + \frac{|R_2|}{|S|} \cdot \varepsilon_1(R_2)\right) = \frac{1}{3} - \left(\frac{1}{3} \cdot 0 + \frac{2}{3} \cdot \frac{1}{2}\right) = 0.$$

- Test T_3 gives the split

$$L_3 = \{x_2, x_3\}, \quad R_1 = \{x_1\}.$$

The corresponding gain is

$$G_3(S) = \varepsilon_1(S) - \left(\frac{|L_3|}{|S|} \cdot \varepsilon_1(L_3) + \frac{|R_3|}{|S|} \cdot \varepsilon_1(R_3)\right) = \frac{1}{3} - \left(\frac{2}{3} \cdot \frac{1}{2} + \frac{1}{3} \cdot 0\right) = 0.$$

We choose the splitting e.g. by the test T_2. The data subset L_2 ends up in a leaf C_{yes}. Finally, we split $S = R_2$ by the remaining test T_3:

$$L_3 = \{x_2\}, \quad R_3 = \{x_1\}.$$

The data subset L_3 ends up in a leaf C_{no}, and R_3 in C_{yes}. Overall, we obtain the decision tree D as depicted in Fig. 10.10. For its average external path length we have:

$$\rho(D) = \frac{1 + 2 + 3 + 3}{4} = \frac{9}{4}.$$

Fig. 10.10 Suboptimal
decision tree D

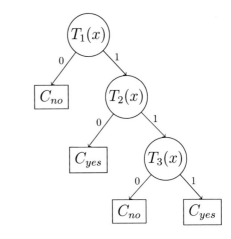

Fig. 10.11 Optimal decision
tree $\bar{D}$

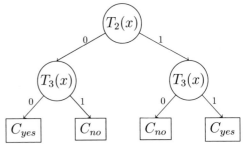

The minimal average external path length is achieved by the decision tree $\bar{D}$, see
Fig. 10.11:

$$\rho\left(\bar{D}\right) = \frac{2+2+2+2}{4} = 2.$$

Note that both D and $\bar{D}$ have zero misclassification rate on the data set $\mathcal{X}$.

Exercise 9.5 (Information Gain)
We recall the definition of the gain with respect to the entropy:

$$G(S) = \varepsilon_2(S) - \left(\frac{|L|}{|S|} \cdot \varepsilon_2\left(L\right) + \frac{|R|}{|S|} \cdot \varepsilon_2\left(R\right)\right).$$

By substituting, we have:

$$G(S) = -p_S \cdot \log_2 p_S - (1 - p_S) \cdot \log_2 (1 - p_S)$$

$$-q_S \cdot \left(-p_L \cdot \log_2 p_L - (1 - p_L) \cdot \log_2 (1 - p_L) \right)$$

$$- (1 - q_S) \cdot \left(-p_R \cdot \log_2 p_R - (1 - p_R) \cdot \log_2 (1 - p_R) \right).$$

Note that it holds:

$$H(X) = -p_S \cdot \log_2 p_S - (1 - p_S) \cdot \log_2 (1 - p_S),$$

$$H(Y) = -q_S \cdot \log_2 q_S - (1 - q_S) \cdot \log_2 (1 - q_S).$$

Moreover, the joint probability distribution of the random variable (X, Y) is

$$q_S \cdot p_L, \quad q_S \cdot (1 - p_L), \quad (1 - q_S) \cdot p_R, \quad (1 - q_S) \cdot (1 - p_R).$$

Then, we have:

$$H(X, Y) = -q_S \cdot p_L \cdot \log_2 q_S \cdot p_L$$

$$-q_S \cdot (1 - p_L) \cdot \log_2 q_S \cdot (1 - p_L)$$

$$- (1 - q_S) \cdot p_R \cdot \log_2 (1 - q_S) \cdot p_R$$

$$- (1 - q_S) \cdot (1 - p_R) \cdot \log_2 (1 - q_S) \cdot (1 - p_R).$$

Altogether, we obtain:

$$G(S) = H(X) + H(Y) - H(X, Y).$$

Exercise 9.6 (ID Decision Tree)
We start to construct the decision tree by applying the iterative dichotomizer with the entropy ε_2 as generalization error. At the beginning, the set S contains 14 elements. Besides, at the first node, there are the three features fever, cough and dyspnea, where we can choose tests from. Let us compare the gains from the corresponding splits. First, we calculate the entropy

$$\varepsilon_2(S) = -p_S \cdot \log_2 p_S - (1 - p_S) \cdot \log_2 (1 - p_S),$$

where the probability is given by

$$ps = \frac{|S \cap C_{yes}|}{|S|} = \frac{7}{14}.$$

Thus, the class labels are equally distributed and, hence, $\varepsilon_2(S) = 1$. In order to decide which test T_j, $j = 1, 2, 3$, to choose, we have to apply the splitting rule by choosing

$$j \in \arg \max_{j \in \{1,2,3\}} G_j(S).$$

We recall that each test T_j splits the data into subsets

$$L_j = \{x_i \in S \mid T_j(x_i) = 0\}, \quad R_j = \{x_i \in S \mid T_j(x_i) = 1\}.$$

The gain associated with this split is

$$G_j(S) = \varepsilon_2(S) - \left(\frac{|L_j|}{|S|} \cdot \varepsilon_2(L_j) + \frac{|R_j|}{|S|} \cdot \varepsilon_2(R_j) \right).$$

Let us compute these gains. We start with a split by fever and denote the corresponding test T_1. This yields the subsets

$$L_1 = \{x_1, x_6, x_9, x_{11}, x_{12}, x_{13}\}, \quad R_1 = \{x_2, x_3, x_4, x_5, x_7, x_8, x_{10}, x_{14}\}.$$

Note that in the set L_1 only person x_9 is infected, i.e. $p_{L_1} = \frac{|L_1 \cap C_{yes}|}{|L_1|} = \frac{1}{6}$. Hence, we calculate the entropy $\varepsilon_2(L_1) = 0.65$ by rounding. Within the set R_2 two persons x_3 and x_{14} are labeled as not infected, hence, $p_{R_1} = \frac{|R_1 \cap C_{yes}|}{|R_1|} = \frac{6}{8}$. From this we compute the entropy $\varepsilon_2(R_1) = 0.81$. We are ready to compute the associated gain:

$$G_1(S) = 1 - \left(\frac{6}{14} \cdot 0.65 + \frac{8}{14} \cdot 0.81 \right) \approx 0.256.$$

Analogously, let us also present the results for T_2, the feature cough, in detail. A split according to cough provides the sets

$$L_2 = \{x_1, x_4, x_7, x_8\}, \quad R_2 = \{x_2, x_3, x_5, x_6, x_9, x_{10}, x_{11}, x_{12}, x_{13}, x_{14}\}.$$

There are three infected persons in the set L_2, hence, $p_{L_2} = \frac{3}{4}$ and $\varepsilon(L_2) = 0.81$. There are four samples in R_2, which are classified as infected, and we obtain $p_{R_2} = \frac{2}{5}$, as well as $\varepsilon_2(R_2) = 0.97$. Therefore, the gain from the split associated with the feature cough is

$$G_2(S) = 1 - \left(\frac{4}{14} \cdot 0.81 + \frac{10}{14} \cdot 0.97 \right) \approx 0.076.$$

We see, that by choosing test T_1 we would gain more information, i.e. the symptom fever distinguishes infected persons from not infected better than cough. It remains to calculate the information gain of T_3. The persons without and with dyspnea are

$$L_3 = \{x_1, x_3, x_6, x_{10}, x_{11}, x_{14}\}, R_3 = \{x_2, x_4, x_5, x_7, x_8, x_9, x_{12}, x_{13}\}.$$

Within L_3 only person x_{11} is infected. Further, six persons from R_3 are classified as infected. Thus, we obtain the same information gain as for the test T_1:

$$G_3(S) = 1 - \left(\frac{6}{14} \cdot 0.65 + \frac{8}{14} \cdot 0.81 \right) \approx 0.256.$$

The greedy algorithm is looking for the maximum gain at each node, so we could choose either T_1 or T_3. We proceed by splitting according to fever. On the left subtree, we have the remaining set

$$L_1 = \{x_1, x_6, x_9, x_{11}, x_{12}, x_{13}\}.$$

The algorithm now sets $S = L_1$ and starts from the beginning, as neither the set is empty nor all instances of the set have the same label. The remaining tests at this subtree are T_2 and T_3. The former would give the singleton set $L_2 = \{x_1\}$ with zero entropy. In R_2 there are five persons with one of them being infected. Hence, $p_{R_2} = \frac{1}{5}$ and $\varepsilon_2(R_2) = 0.72$. Altogether the gain at the node of this subtree is

$$G_2(S) = 0.65 - \left(\frac{1}{6} \cdot 0 + \frac{5}{6} \cdot 0.72 \right) \approx 0.05.$$

We compare it to the gain associated with T_3. The corresponding subsets are

$$L_3 = \{x_1, x_6, x_{11}\}, \quad R_3 = \{x_9, x_{12}, x_{13}\},$$

and the resulting entropies are

$$\varepsilon_2(L_3) = 0, \quad \varepsilon_2(R_3) = 0.92.$$

Hence, the information gain is $G_3(S) = 0.19$, which is clearly higher than $G_2(S)$. We therefore split by using dyspnea, and the node associated to L_3 becomes a leaf with label C_{no}. On the right subtree the algorithm sets $S = R_1$, and we also have the choice between two splits T_2 and T_3. The former would yield the information gain

$$G_2(S) \approx 0.20.$$

T_3 provides the sets

$$L_3 = \{x_3, x_{10}, x_{14}\}, \quad R_3 = \{x_2, x_4, x_5, x_7, x_8\}.$$

All persons in R_3 are infected, hence we get a leaf with label C_{yes}. The information gain is higher than for T_2, as

$$G_3(S) = 0.81 - \left(\frac{3}{8} \cdot 0.92 + \frac{5}{8} \cdot 0 \right) \approx 0.465.$$

Thus, we choose to use the information provided by dyspnea. We present the temporary decision tree in Fig. 10.12. We proceed on the left subtree with the edge labeled by 1 and outgoing from T_3. The algorithm sets $S = R_3$ and uses the only remaining test on cough. As all of the three persons have this symptom, L_3 is empty, and all persons are in R_3. One of them is infected, which provides $p_{R_3} = \frac{1}{3}$. After this, the set of tests is empty, so the algorithm creates a leaf by the majority vote. Hence, we get a leaf with label C_{no}. The empty subset L_3 can be labeled by either C_{yes} or C_{no}. The situation on the right subtree is similar. Again, all remaining persons suffer from cough. The application of the majority vote to R_3 once more creates the leaf C_{no}. The empty subset L_3 can be labeled by either C_{yes} or C_{no}. Let us have a look on the final ID decision tree D, see Fig. 10.13.

Let us elaborate how we can prune the decision tree D. Note that it achieves the misclassification rate $\mu(D) = \frac{2}{14}$. The feature cough provides no new information. If we replace this test on the left by the leaf C_{no}, the misclassification rate remains unchanged. But, after applying T_3 both sets would be classified as not infected, so we could also prune

Fig. 10.12 Temporary
decision tree

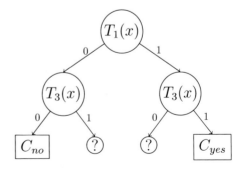

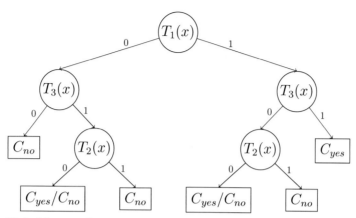

Fig. 10.13 ID decision tree D

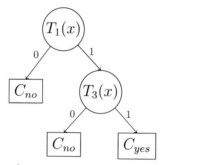

Fig. 10.14 Pruned decision tree D'

this node to a leaf C_{no}. On the pruned left subtree only person x_9 is then misclassified, but even the initial decision tree D is not able to classify person x_9 as infected. On the right subtree we can prune the node T_2 and use the leaf C_{no} instead, by the same argument as before. Any further pruning on the right subtree would increase the misclassification rate. We are ready to present the pruned decision tree D' in Fig. 10.14. According to the latter, the features fever and dyspnea are very important symptoms in order to detect infection. Note that a person must have both these symptoms in order to be classified as infected. On the contrary, the symptom cough can be neglected. E.g., a person without fever and cough, but with dyspnea is predicted to be not infected.

Bibliography

C.C. Aggarwal, *Recommender Systems* (Springer, New York, 2016)

D. Arthur, S. Vassilvitskii, *k*-means++: the advantages of careful seeding, in *Proceedings of the eighteenth annual ACM-SIAM symposium on Discrete algorithms*, 2007, pp. 1027–1035

L. Ball, D. Leigh, P. Loungani, Okun's law: fit at 50? Journal of Money, Credit and Banking **59**, 1413–1441 (2017)

A. Beck, M. Teboulle, Mirror descent and nonlinear projected subgradient methods for convex optimization. Operations Research Letters **31**, 167–175 (2003)

A. Beck, M. Teboulle, A fast iterative shrinkage-thresholding algorithm for linear inverse problems. SIAM Journal of Image Sciences **2**, 183–202 (2009)

L. Breiman, J. Friedman, R. Olshen, C. Stone, *Classification and Regression Trees* (Brooks/Cole Publishing, Monterey, 1984)

S. Brin, L. Page, The anatomy of a large-scale hypertextual Web search engine. Computer Networks and ISDN Systems **30**, 107–117 (1998)

S.S. Chen, D.L. Donoho, M.A. Saunders, Atomic decomposition by basis pursuit. SIAM Journal on Scientific Computing **20**, 33–61 (1998)

C. Cortes, V.N. Vapnik, Support-vector networks. Machine Learning **20**, 273–297 (1995)

G. Cybenko, Approximations by superpositions of sigmoidal functions. Mathematics of Control, Signals, and Systems **2**, 303–314 (1989)

J.R. Firth, *A Synopsis of Linguistic Theory 1930–1955*. Studies in Linguistic Analysis (Longman, Green & Co., London, 1957)

R.A. Fisher, The use of multiple measurements in taxonomic problems. Annals of Eugenics **7**, 179–188 (1936)

S. Foucart, H. Rauhut, *A Mathematical Introduction to Compressive Sensing* (Springer, New York, 2013)

D. Gale, *The Theory of Linear Economic Models* (University of Chicago Press, Chicago, 1960)

S. Haykin, *Neural Networks and Learning Machines* (Pearson Education, New York, 2011)

E. Hazan, *Introduction to Online Convex Optimization* (Now Publishers Inc, Hanover, US, 2016)

D.F. Hendry, B. Nielsen, *Econometric Modeling: A Likelihood Approach* (Princeton University Press, Princeton, New Jersey, 2014)

K. Hornik, Approximation capabilities of multilayer feedforward networks. Neural Networks **4**, 251–257 (1991)

C.-J. Hsieh, K.-W. Chang, C.-J. Lin, S. Keerthi, S. Sellamanickam, A dual coordinate descent method for large-scale linear SVM, in *Proceedings of the 25th International Conference on Machine Learning*, 2008

G.F. Hughes, On the mean accuracy of statistical pattern recognizers. IEEE Transactions on Information Theory **14**, 55–63 (1968)

L. Hyafil, R.L. Rivest, Constructing optimal binary decision trees is NP-complete. Information Processing Letters **5**, 15–17 (2009)

P. Jaccard, Lois de distribution florale dans la zone alpine. Bulletin de la Société Vaudoise des Sciences Naturelles **38**, 70–130 (1902)

H.T. Jongen, K. Meer, E. Triesch, *Optimization Theory* (Kluwer Academic Publishers, Boston, 2004)

E.R. Kandel, *Principles of Neural Science* (McGraw-Hill Medical, New York, 2013)

R.M. Karp, Reducibility among combinatorial problems, in *Complexity of Computer Computations. Proceedings of a Symposium on the Complexity of Computer Computations*, ed. by R.E. Miller, J.W. Thatcher (Plenum, New York, 1972), pp. 85–1103

S.Y. Kung, *Kernel Methods and Machine Learning* (Cambridge University Press, New York, 2014)

P. Lancaster, *Theory of Matrices* (Academic Press, London, 1969)

V.I. Levenshtein, Binary codes capable of correcting deletions, insertions, and reversals. Soviet Physics Doklady **10**, 707–710 (1966)

S.P. Lloyd, Least square quantization in PCM. IEEE Transactions on Information Theory **28**, 129–137 (1982)

H.M. Markowitz, Portfolio selection. The Journal of Finance **7**, 77–91 (1952)

R. Mathar, G. Alirezaei, E. Balda, A. Behboodi, Fundamentals of Data Analytics (Springer, Cham, Switzerland, 2020)

O.A. Mcbryan, GENVL and WWWW: Tools for taming the Web, in *Proceedings of the first World Wide Web Conference*, 1994

K. Murphy, *Machine Learning: A Probabilistic Perspective* (The MIT Press, Cambridge, Massachusetts, 2012)

Yu. Nesterov, *Lectures on Convex Optimization* (Springer, Cham, Switzerland, 2018)

Yu. Nesterov, A. Nemirovski, Finding the stationary states of Markov chains by iterative methods. Applied Mathematics and Computation **255**, 58–65 (2015)

J.R. Quinlan, Induction of decision trees. Machine Learning **1**, 81–106 (1986)

R.T. Rockafellar, *Convex Analysis* (Princeton University Press, Princeton, New Jersey, 1970)

L. Rokach, O. Maimon, *Data Mining with Decision Trees: Theory and Applications* (World Scientific, Singapore, 2015)

F. Rosenblatt, The Perceptron - a perceiving and recognizing automaton, Report 85–460–1, Cornell Aeronautical Laboratory, 1957

M. Sahlgren, The distributional hypothesis. Rivista di Linguistica **20**, 33–53 (2008)

P.A. Samuelson, W.D. Nordhaus, *Economics* (McGraw-Hill, New York, 2004)

P.A. Samuelson, T.C. Koopmans, J.R.N. Stone, Report of the evaluative committee for econometrica. Econometrica **22**, 141–146 (1954)

S. Shalev-Shwartz, S. Ben-David, *Understanding Machine Learning* (Cambridge University Press, New York, 2014)

W.F. Sharpe, Capital asset prices: a theory of market equilibrium under conditions of risk. The Journal of Finance **19**, 425–442 (1964)

H. Steinhaus, Sur la division des corps matériels en parties. Bull. Acad. Polon. Sci. (in French) **14**, 801–804 (1957)

R. Tibshirani, Regression shrinkage and selection via the lasso. Journal of the Royal Statistical Society **58**, 267–288 (1996)

A.N. Tikhonov, V.Y. Arsenin, *Solution of Ill-posed Problems* (Winston & Sons, Washington, 1977)

A.M. Tillmann, M.E. Pfetsch, The computational complexity of the restricted isometry property, the nullspace property, and related concepts in compressed sensing. IEEE Transactions on Information Theory **60**, 1248–1259 (2013)

V. Vovk, C. Watkins, Universal portfolio selection, in *Proceedings of the Annual ACM Conference on Computational Learning Theory*, 1998, pp. 12–23

H. Zellig, Distributional structure. Word **10**, 146–162 (1954)

M. Zinkevich, Online convex programming and generalized infinitesimal gradient ascent, in *Proceedings of the 20th International Conference on Machine Learning*, 2003, pp. 928–936

Index

© Springer-Verlag GmbH Germany, part of Springer Nature 2021
V. Shikhman, D. Müller, *Mathematical Foundations of Big Data Analytics*,
https://doi.org/10.1007/978-3-662-62521-7

Printed in the United States
By Bookmasters